XAVIER BICHAT AND THE MEDICAL THEORY OF THE EIGHTEENTH CENTURY

(*Medical History*, Supplement No. 4)

XAVIER BICHAT AND THE MEDICAL THEORY OF THE EIGHTEENTH CENTURY

by

ELIZABETH HAIGH

(*Medical History*, Supplement No. 4)

LONDON
WELLCOME INSTITUTE FOR THE HISTORY OF MEDICINE
1984

Published 1984 by the Wellcome Institute for the History of Medicine, 183 Euston Road, London NW1 2BP, England.

ISBN 0 85484 046 X
ISSN 0025 7273 4

Supplements to *Medical History* may be obtained at the Wellcome Institute, or by post from Science History Publications Ltd, Halfpenny Furze, Mill Lane, Chalfont St Giles, Bucks HP8 4NR, England.

Printed by the Wellcome Foundation Limited, Print and Packaging Division (Crewe).

CONTENTS

PREFACE

This work grew from an observation made when I was a graduate student at the University of Wisconsin. Reading about nineteenth-century medical giants, I frequently came across passing references to Xavier Bichat, reputedly a major vitalist and the father of histology. It seemed clear that at the beginning of the last century, the French medical establishment revered his contributions to medical teaching and to anatomy and physiology. But what had been said about him more recently, albeit flattering, tended to be vague, as though no one was quite sure what all the fuss had been about. A question to Dr Nikolaus Mani about why Bichat was so little known was the occasion for him suggesting Bichat's work as a suitable dissertation topic. That was completed more than a decade ago. Thereafter, I moved a considerable distance from Bichat into the earlier part of the eighteenth century. But as I read the work of many authors and especially vitalists of various sorts, I continually encountered foreshadowings of what Bichat wrote later. Thus this project, which studies Bichat in the context of eighteenth-century traditions, came to be.

In the meantime, other persons have contributed much material to the study of the medical, physiological, and social developments in the latter half of the eighteenth century, thereby elucidating Bichat's considerable contributions to developments in post-revolutionary France.

Persons who have read the manuscript at various stages of its preparation and to whom I owe special thanks for their advice and observations are the late Walter Pagel, Roger French, L. Pearce Williams, Frank Dougherty, and William Bynum.

I am grateful to the librarians and other staff members of the following centres for their unfailing patience and courtesy: the Memorial Library of the University of Wisconsin in Madison, the Bibliothèque Nationale and the Bibliothèque de l'Ecole de Médecine in Paris, the Bibliothèque de l'Université de Montpellier, Cambridge University Library, and the Library of the Wellcome Institute for the History of Medicine in London.

I acknowledge two research grants from the Canada Council, now the SSHRC. A travel grant allowed me to go to Montpellier in the summer of 1974, and a Leave Fellowship spared me the responsibility of teaching in the 1975–76 academic year. A Wellcome Fellowship awarded for 1977–78 permitted me to spend an enriching year at the University of Cambridge. That same year, a British Council Commonwealth University Interchange Scheme Travel Grant paid the expenses for travel to the United Kingdom.

Finally, I thank my husband David for unfailing support and patience.

I

THE LIFE OF A YOUNG PHYSICIAN

The only statue which stands in the courtyard of the Medical Faculty of Paris is that of a young man who neither attended nor taught in a formal medical school. The figure was erected in 1857 to honour the memory of Marie-François-Xavier Bichat, who had died fifty-five years earlier. Along with many young Frenchmen during the revolutionary years, Bichat obtained his medical education as an apprentice to senior men. For a time, he assisted medical practitioners who served with the French armies in the field. He went to Paris in 1794, less than a year after the revolutionary tribunal had decreed the abolution of all university faculties. The lack of a formal education did not hinder Bichat particularly. Though he was only thirty years old when he died, he had accomplished a formidable quantity of work which earned him a considerable fame, at least in Paris. Within the next few years, his reputation spread far throughout Europe and even into North America.

Bichat was known first as a spokesman for a vitalist theory of the life sciences and second as the creator of the tissue theory of anatomy. As a vitalist, he argued eloquently that life sciences are different from physical sciences because living matter is fundamentally different in its properties and its behaviour from that which is inert. His arguments influenced physiological theory well into the nineteenth century. When François Magendie and Claude Bernard argued in the nineteenth century that physiologists erred in assuming that there were two distinct natural realms, they were speaking largely to Bichat's position. Bichat located what he called "vital properties", which he believed to be unique to living matter, in the body's tissues. He isolated twenty-one tissues in the body and he showed that they unite in various combinations to produce its organs and structures much as chemical elements produce compounds. This tissue theory of anatomy became the basis of the science of histology and it has often been judged to be Bichat's most important contribution to medical and physiological theory.

During his brief career as a physician, physiologist, and anatomist, Bichat taught a considerable number of students. They apparently held him in warm regard as a man who could inspire them with his own enthusiasm for his subject. His associates at Paris's largest hospital, the Hôtel-Dieu, and his colleagues in various professional societies described Bichat as a modest, warm man of a gentle and open nature. Such affirmations and recollections about the dead are common enough, of course. Bichat's funeral, however, suggests that his various biographers were not merely offering token tributes. It was a moving occasion, fifty carriages draped in black accompanying the body in the procession. On hearing of Bichat's death, the First Consul Napoleon Bonaparte instructed his Minister of the Interior to place a marble monument in the Hôtel-Dieu to honour Bichat and his teacher, the surgeon Pierre Desault. Since then, in their inimitable way, the French have continued to pay tribute to Bichat's memory with numerous plaques, paintings, statues, busts, engravings, odes, eulogies, and biographies. He has been the subject of innumerable student theses. In Paris, a street and hospital are named after Bichat; he is represented on the pediment of the Pantheon; the Faculty of Medicine has a statue of him inside the building as well as one in the courtyard.

Xavier Bichat was born in 1771, the first of four children, in a small fishing village called Thoirette-en-Bas near Lyons.[1] This was the family home of his parents, who were officially classified as bourgeois. The boy was brought up nearby in a small town called Poncin, where his father Jean-Baptiste practised medicine, having received his medical degree from the University of Montpellier in 1769. In the middle of the eighteenth century, most of the medical faculty of that ancient medical school adhered to a vitalist theory of living matter. One of the elder Bichat's teachers had been the vitalist Paul-Joseph Barthez, whose influence and prestige in Montpellier were unparalleled in the eighteenth century. Barthez' ideas about the nature of the life sciences and their distinction from physical ones were to influence Xavier Bichat greatly in subsequent years. It may be that his father's work sparked an interest in the life sciences in his first-born son. Some years after Xavier's death, his younger brother César described him as having had a precocious interest in science. By the age of seven, Xavier was purportedly practising dissection on the cats and dogs of the neighbourhood. He liked to hunt, but only when he could dissect his trophies. I expect that those enthusiastic recollections were merely exaggerations to suit an occasion. If not, it is possible that today Bichat would have been sent to reform school.

At the age of eleven, like his father before him, Bichat was sent to the Jesuit Collège de Nantua near Lyons. His parents withdrew him from there in August 1790 because they did not share the enthusiasm of the townsfolk and even of the school officials for the Revolution, which had begun the previous year. From Nantua, Bichat went to the Seminary of Saint Irénée in Lyons, where one of his uncles was Superior. The Revolution and France's wars with her neighbours were henceforth to intrude themselves frequently into the young man's life. Had the political upheaval not occurred, Bichat

[1] There are many papers and theses which outline Bichat's life and work. The most recent and accurate account is by Maurice Genty, 'Xavier Bichat (1771–1802)', in Pierre Huard (editor), *Biographies médicales et scientifiques*, Paris, Dacosta, 1972. An earlier one by the same author is 'Bichat et son temps', *La médecine internationale illustrée*, serialized monthly from July 1934 to September 1935. It deals with Bichat's family, his personal and professional life, and even with the lives of his major associates. His contributions to medicine and physiology are not treated in any detail. Genty also published a six-page pamphlet, 'Quelques documents sur la famille et la maison natale de Bichat', *Bull. Soc. Hist. Med.*, 1933, **28**: nos. 3–4. He found the first record of the Bichat family in the 1682 parochial archives of Thoirette. Jacques Coquerelle had published a short book *Xavier Bichat, (1771–1802)*, Paris, A. Maloine, 1902, the first part of which is a profusion of details about the family, its genealogy, personal and professional letters, and a brief survey of Bichat's army and medical careers. Coquerelle did painstaking archival work in Thoirette, the Collège de Nantua, and in Paris. A large part of his book is given over to reproducing eulogies and odes delivered in Bichat's honour. A diverse collection of material was put together by Prof. Raphael Blanchard in *Centenaire de la mort de Xavier Bichat*, Paris, Librairie Scientifique et Littéraire, 1903. It contains two biographical speeches which Blanchard delivered on the occasion of the centenary celebrations, one at Bichat's tomb and the other before his house in the rue des Carmes. A section of 'Documents inedits concernant Xavier Bichat' contains letters and military and professional documents. Finally, 'Documents artistiques relatifs à Xavier Bichat' gives a list of portraits, plaques, busts, statues, and medals that honour him. A few unique and personal features are found in a biographical sketch by Bichat's cousin Mathieu-François Buisson, 'Précis historique sur Marie-François-Xavier Bichat', *Traité d'anatomie descriptive*, 5 vols., Paris, Brosson, Gabon, 1802, vol. 3, pp. vii-xxviii. Unless otherwise indicated, the biographical material contained in Chapter 1 is taken from these various sources.

would have studied anatomy and surgery as an apprentice and then perhaps have taken medicine at Montpellier. He might even have returned to a medical practice in the pastoral countryside of the Rhône Valley. It was not to be, however.

The chief hospital in a French city or town is called the Hôtel-Dieu. Bichat began courses in anatomy and surgery at the Hôtel-Dieu in Lyons in 1791. It was only in the eighteenth century that the status of surgery had been raised sufficiently to separate surgeons from their long-standing association with the company of barbers. In spite of vociferous protests, especially from the powerful Medical Faculty of Paris, which traditionally blocked virtually all innovation, a Royal Academy of Surgery was created in 1731 and empowered to grant a Master of Surgery degree. It was part of an attempt to put surgeons on a level with the considerably more prestigious physicians, who alone among medical practitioners were university trained. In the 1790s, most students still learned surgical skills by attaching themselves as apprentices to a master. Since 1788, the Chief Surgeon at the Hôtel-Dieu in Lyons was Marc-Antoine Petit. For two years, he had been a student of Pierre-Joseph Desault, the Chief Surgeon of the Hôtel-Dieu in Paris between 1785 and 1795, and the pre-eminent surgeon in France.[2] Desault was to become Bichat's master in Paris in 1794. Europe's hospitals at the end of the eighteenth century were filthy, vermin-ridden structures in which at least as much disease was contracted as cured. Lacking ventilation, privacy, and even minimal hygiene, the institutions mixed sufferers from smallpox and syphilis among expectant mothers. People scheduled for surgery listened to the screams of those undergoing it. Mortality was high. In the Hôtel-Dieu of Paris it was about one in every three patients; that of Lyons seems to have been somewhat better, having modernized to the point where each of its possible 400 patients could have his own bed. In Paris, as many as four patients might occupy the same bed, especially when the hospital was crowded in unhealthy seasons and during epidemics.[3]

The year that Bichat began his surgical training was a difficult one for all the learned professions. The institutions of the former regime were increasingly under attack as élitist structures which must be purged of their privileged membership or destroyed altogether. The theory and outlines of a new educational system were being widely discussed, but it was to be years before a new structure replaced the old one. In the meantime, the members of the National Assembly were effectively creating a kind of educational vacuum as they set about destroying the symbols of former privilege. In a flurry of egalitarian sentiment, the National Assembly abolished the master's degree in medicine as well as all examinations for medical degrees in February 1791. In March, it decreed that anyone paying a fee could be licensed to practise medicine and surgery without examination. At about the same time, some members were purged from the Academy of Medicine.

Such steps exacerbated the health crisis that followed France's declaration of war

[2] The growth of the professional status of surgery in France, and its role in the new medicine of post-revolutionary France are discussed in Toby Gelfand, *Professionalizing modern medicine: Paris surgeons and medical science and institutions in the eighteenth century*, Westport, Conn., and London, Greenwood Press, 1980.

[3] This is discussed by David M. Vess in *Medical revolution in France, 1789–1796*, Gainesville, University Press of Florida, 1975, pp.10–39. The Hôtel-Dieu in Paris is described by Charles Coulton Gillispie, *Science and polity in France at the end of the old regime*, Princeton University Press, 1980, pp. 244–256.

on Austria and Prussia in April 1792. The Legislative Assembly authorized the drafting of medical personnel and the requisitioning of private houses, chateaux, convents, and churches for military hospitals.[4] One such was established almost immediately in the former Seminary of Saint Irénée. Bichat had to interrupt his studies in November 1792 in order to serve as a *chirurgien-surnuméraire* in his former school, a kind of junior assistant to the qualified surgeons.

In February 1793, the Vendée rose in revolt in reaction to conscription, the Revolution, and the rigours of war in general. The revolt grew into a civil war which threatened to tear France apart. To deal with the combined internal and external crises, the National Assembly in April 1793 created the Committee of Public Safety headed by Georges Danton. The Committee decreed that the army must first suppress the revolt. On 29 April, during the ensuing fight, the Jacobins lost control of the city of Marseilles. Lyons broke out into full-scale revolt with severe street fighting a month later. Bichat participated in this attempt on the part of royalists and moderates to overthrow the Jacobin municipal authorities. For a brief time they were successful.

On 2 June 1793, Jean-Paul Marat's supporters in Paris expelled the more moderate Girondin deputies from the Convention, leaving unquestioned political authority in Jacobin hands. The arrest of twenty-nine Girondin deputies precipitated another reaction outside the capital. In Lyons, Marseilles, and Toulon, Jacobins were guillotined or hanged. In July, part of the Alpine Army was ordered to march against Lyons, Bichat helped to prepare the city's defences. But when a decree ordered all persons who were not citizens to leave Lyons on pain of being considered conspirators, Bichat left to return home to Poncin.

On 1 August, the same day Bichat left Lyons, the Ministry of War began a requisition of all health officers, physicians, surgeons, and apothecaries between the ages of eighteen and forty years to be divided among the armies of the Republic. It was a time of grave crisis and there was little reference made to ability or training. Hospitals were designated as teaching centres in which the more qualified medical practitioners would instruct those who had not completed their formal training. Bichat was drafted and sent to Grenoble and then to Bourg to serve as a surgeon in the Department of the Ain whose hospitals were put under military control. Wounded soldiers from the troops besieging Lyons were sent there. At the end of 1793, the hospital was dissolved. Again, Bichat returned to Poncin.

At home, he found that the Revolution had provoked a predictable resentment against the bourgeois Bichat family. In the early stages of the Revolution, Jean-Baptiste Bichat had been a deputy of the Third Estate. More recently, however, he was accused by the sansculottes of a lack of patriotism. It was a common enough accusation and a sufficiently vague one to be very widely applicable. The accusation also touched the sons of the family. In order to persuade their detractors that they were indeed patriotic, Xavier and César enrolled in work for military ambulances and voluntary battalions respectively. Because of the local pressures, Xavier left for Paris in June 1794. For a time, he lived with his uncle, aunt, and cousin Buisson, who had fled to the capital from Lyons during the siege. It was an invaluable move, for there Bichat gained access to the sort of professional and intellectual society which Paris

[4] Vess, op cit., note 3 above, pp. 40–53.

traditionally drew to itself, and he was at last permitted to study seriously and to practise medicine.

When the Revolution started in 1789, there were eighteen medical schools in France, although only Paris and Montpellier enjoyed any particular reputation. There were also fifteen colleges of surgery and a college of pharmacy. Few physicians would have disputed the need for a reform of medical education, but the revolutionaries, especially the Girondists, were unwilling to permit the profession to reform itself. They exaggerated the abuses of many professions, accusing them in general of being the preserve often of charlatans. The Convention had already meddled in medical education in 1791, as we have seen. In 1793, its members went even further. On 28 August, the Convention abolished all literary academies, the Royal Academy of Medicine, and the Royal Academy of Surgery. On 15 September, all the university faculties in France were declared closed.[5] Thus the revolutionary republicans had carried to its logical conclusion their conviction that the savants of the old regime and their aristocratic institutions were an affront to their Revolution's egalitarian goals. With the closure of these bastions of aristocracy, idealists hoped, as idealists will, that henceforth there would be no examinations and no qualifications other than age, experience, and respect for citizens. Whoever wished to teach mathematics, law, or medicine would have only to obtain from his municipality a certificate of integrity and good citizenship. It was assumed that freely given lessons would be largely paid for by the students. In the meantime, other persons persisted in wanting to resurrect a reformed system of education. Accordingly, the Convention urged the Committee of Public Education, of which the chemist Antoine Fourcroy was an important member, to hurry a report on a proposed reorganization of the system.

By the autumn of 1793 and throughout 1794, the only course of action open to men such as Bichat who were seeking medical instruction was to attach themselves to hospitals as apprentices. Desault, the founder of clinical teaching of surgery in France attracted many such private students, including Bichat, as well as those attached to hospitals. He had been a member of the Royal Academy of Surgery and, since 1786, Chief Surgeon of the Hôtel-Dieu. The position involved treatment of some 400 patients. In 1791, Desault was responsible for ninety non-resident surgical students or *externes* and for fourteen resident *internes*. In an amphitheatre he set up in 1788, he offered surgical demonstrations which were attended also by the private students. By 1791, Desault was providing daily anatomical dissection demonstrations before as many as 300 students.[6] In May 1793, just when the civil war crisis was developing, Desault was arrested and brought to trial before a Revolutionary Committee, charged

[5] Theodore Puschmann, *A history of medical education*, New York, Hafner, 1966, facsimile of 1891 edition, pp. 420–421. The institution of clinical hospital courses in surgery in Paris is discussed by Toby Gelfand, 'The hospice of the Paris College of Surgery (1774–1793); a unique and invaluable institution', *Bull. Hist. Med.*, 1973, **47**: 375–393. The attitudes towards hospitals and towards medical instruction during the revolutionary period are discussed in Michel Foucault, *The birth of the clinic*, trans. by A. M. Sheridan, London, Tavistock, 1973, pp. 38–53.

[6] Gelfand, op. cit., note 2 above, pp. 116–125, discusses Desault specifically. See also Toby Gelfand, 'A confrontation over clinical instruction at the Hôtel-Dieu of Paris during the French Revolution', *J. Hist. Med.*, 1973, **28**: 268–282. This article relates the circumstances of a complaint from a delegation of "citizen students of surgery" to the Committee of Public Instruction. Gelfand translates a letter from Desault to that Committee composed in November 1791. It relates a good deal about his teaching methods.

with refusing aid to the wounded, but he was acquitted in August, due largely to the efforts of Antoine Fourcroy.

The Terror was at its height when Bichat arrived in Paris in June 1794. Though the Revolutionary Tribunal had already reaped a harvest of heads, executions multiplied after 10 June. That day, in order to destroy the Girondist Georges Danton, Robespierre induced the Tribunal to pass the *Loi de prairial*, which denied a defence at the trial of anyone accused of an offence against the state. On 28 July, the 9th of Thermidor according to the revolutionary calendar, the guillotine finally decapitated Robespierre himself and put an end thereby to the excessive Terror which had shaken France. Thereafter, many physicians and other citizens who had been suspended for disloyalty since September 1791 were recalled to active duty with the troops in the field. I have found no reference to political events in any extant work or letter composed by Bichat.[7] He assiduously avoided any direct political involvement, even though he was thrown into association with many intellectuals who were variously involved in the re-evaluation and reform of French institutions.

Bichat's natural abilities were reinforced by a remarkable capacity for hard work and self-discipline. Those qualities, along with a stroke of good luck, brought him to Desault's attention. One day, he offered to deliver a public lesson on the subject of a fractured clavicle when the student to whom the task had been assigned was absent. It must have been an impressive performance, for in a letter to his family dated 1 October 1794, he announced that he was living in Desault's home and that he planned to stay in Paris a long time.[8] This arrangement permitted Bichat to complete his education in surgery and anatomy under the tutelage of France's pre-eminent surgeon. It must also have facilitated his social and professional contacts with persons whose ideas would help him to formulate his own notions of anatomy and physiology.

Formal medical training was re-established in France by a decree of 4 December 1794. A major stimulus to do so had apparently been provided in Antoine Fourcroy's report, which noted the death of more than 600 army health officers during the preceding eighteen months and the necessity for recruitment for the army health service. Medical schools were created in Paris, Montpellier, and Strasbourg.[9] The ancient distinction between medicine and surgery was finally abolished for ever, and Desault

[7] Possibly Bichat avoided mentioning politics because to do so would have worried his family. The life of the elder Bichat as well as his reaction to the Revolution are briefly discussed by Jean Rousset, 'Trois docteurs de Montpellier, pères de médecines célèbres nés dans la région Lyonnaise', *Monsp. Hippoc.*, 1965, **8**: 8–11.

[8] Letters from Bichat to his parents are quoted by Emile Jean Kervella in 'La vie et l'oeuvre de Bichat (1771–1802)', a thesis presented to the Paris Medical School for the Doctorat en Médecine in 1931. Among these letters are pleas to his family to understand that he is busy and therefore unable to come home. There are also requests for money.

[9] The Paris Medical School was given twelve professorships: anatomy and physiology, pharmacy, medical physics and hygiene, surgical pathology, pathology of internal diseases, natural history, surgical operations, clinical surgery, clinical medicine, clinical convalescence, obstetrics, history of medicine and forensic medicine. Montpellier was given nine professorships, and Strasbourg six. Each professor was given an assistant. It seems that for a time, the schools were hampered by a shortage of qualified students and had to open their doors to citizens of the French colonies as well as to French nationals. See Puschmann, op. cit., note 5 above, p. 539; Vess, op. cit., note 3 above, pp. 153–184; and Gelfand, op. cit., note 2 above, pp. 149–171. For a discussion of the nature and consequences of the reformed medical education system, and for Bichat's contribution to the reform, see Foucault, op. cit., note 5 above; and Erwin H. Ackerknecht, *Medicine at the Paris Hospital, 1794–1848*, Baltimore, Md., Johns Hopkins University Press, 1967.

was named Professor of Clinical Surgery (*clinique externe*) in the Paris school. His hospital courses at the Hôtel-Dieu became part of the new medical school. Curiously, he and other members of the College of Surgery disapproved of this marriage between the two branches of medical practice, even though it represented the end of a long and bitter battle to raise the professional status of surgery.

Desault died of a fever on 1 June 1795, amid rumours of poison, since he was treating the dying son of Louis XVI. Desault had lived just long enough to permit his special student to establish himself successfully in Paris. Bichat was twenty-four years old when Desault died. He continued to live for most of his remaining six years in the Desault apartment as a kind of adopted son of Madame Desault. We know little of his personal life. The quantity of work which he accomplished suggests that there must have been little time for anything else but the study of anatomy and physiology.

Bichat's training to this point had been primarily surgical, first in Lyons and then in Paris. For a while after Desault's death, his work predictably took its direction from his teacher. In January 1791, Desault had commenced work on a *Journal de chirurgie*, which appeared finally in September 1792. With the help of Corvisart, the holder of the chair of clinical medicine in the Paris Medical School, Bichat collected some of his teacher's unpublished manuscripts and completed a fourth and final volume of the journal in 1795. It included five articles by Bichat, dealing with surgical questions as well as a 'Notice historique sur Desault', his final and public tribute to his esteemed teacher.[10]

Meanwhile, the insecurity created by revolutionary excess was subsiding. On 22 March 1796, members of the former Academies of Medicine and of Surgery, along with other physicians, surgeons, chemists, and pharmacists, formed the Société de Santé de Paris, later renamed the Société de Médecine. Bichat was named a resident member in October 1799. It seems, however, to have been a somewhat stodgy society, and certain professional young men felt the need for another. Bichat was one of the principal founders of the Société Médicale d'Émulation, which met for the first time in the Medical School on 23 June 1796.[11] It was seen to be a kind of successor to the Société Royale de Médecine, which, between 1788 and 1794, attracted to it France's most progressive physicians. "Émulation" is said to have been a word used by Félix Vicq d'Azyr, the perpetual secretary of the Société Royale, to mean professional standards.[12] Clearly, the new group represented the opinions of the new medicine emerging in the post-revolutionary period. The Société d'Émulation's initial resident membership list included Alibert, Cabanis, Fourcroy, Pinel, and Corvisart, and among fifty corresponding members were the names of Bell, Barthez, and Spallanzani. In fact, so many of the most noteworthy members of the Parisian medical profession joined its ranks in its first few years that it became a virtual

[10] Xavier Bichat, 'Eloge de P. J. Desault', *Oeuvres chirurgicales de P. J. Desault*, a new edition corrected and augmented by Bichat, 3 vols., Paris, Mequignon, 1801–03, vol. 1, p. 52.

[11] The Société Médicale d'Émulation deserves more study that it has received so far, since its membership list included the names of many of France's most influential physicians. To date, one of the best studies of that society and especially of the views shared by its members concerning the nature of scientific medicine is found in Sergio Moravia, 'Philosophie et médecine en France à la fin du XVIIIe siècle', *Studies on Voltaire and the eighteenth century*, 1972, **89**: 1089–1151.

[12] Gillispie, op. cit., note 3 above, p. 222.

"Who's Who" of European medicine.

In July 1799, Bichat was invited to join the Société Philomathique de Paris, founded in 1788 to promote its members' knowledge of all the natural sciences.[13] In 1800, he acted as a secretary to the Société de l'École de Médecine de Paris, which was created by the government to provide expert consultants on all questions of health, disease, hygiene, and medical services. His enthusiasm was, nevertheless, always reserved primarily for the Société d'Émulation.

In 1797, Bichat undertook to teach a private anatomy course in a laboratory at 18 rue des Carmes on the Left Bank. His classes appear to have been very successful, this first one attracting nearly eighty students. This sort of teaching, existing alongside that offered by the Medical School, was common at the time. Private teachers, however, faced special problems, for, unlike the Paris Medical School which, in 1795, had 500 cadavers placed at its disposal, they had no access to anatomical material. Bichat resorted to raiding cemeteries. He was arrested in Saint Catherine's cemetery during one such midnight venture, but escaped with a reprimand. The government finally ordered that bodies be made available to dissection laboratories in September 1798. A letter dated 29 November of that year from the Office of Public Health and Welfare authorized Bichat to bury human debris resulting from such dissection. Bichat's teaching operation was expanded in September 1801, when he installed himself at the Collège de Liseaux, also on the rue des Carmes. He both lived and worked in the large apartment for the last ten months of his life.

Bichat directed publication of a collection of Desault's surgical works, which appeared in 1798. Thereafter, his direct involvement with surgery began to give way to an interest in questions of a specifically anatomical and physiological nature. In the second volume of the *Mémoires de la Société Médicale d'Émulation* there are six articles written by Bichat. Three of them deal with surgical topics, but the others provide a preview of the physiological and anatomical work which he was to do thereafter. The first surgical paper was concerned with some modifications of the process of trepanning; the second dealt with fracture of the clavicle, that same lucky bone which had initially brought him to Desault's attention; the third related a new method for ligaturing polyps.[14]

The fourth paper demonstrated that there is a specific organ which actively produces the synovial fluid which lubricates the body joints and articulation.[15] Bichat intended that it be read along with the next paper, in which he divided the bodily membranes into six distinct varieties. He examined their properties and characteristics, treating each one as an anatomical entity in its own right, a specific organic structure which is part of the compound arrangement of organs.[16] A book entitled the *Traité des membranes*, published early in 1800, developed at length the ideas first discussed in these two articles.

Bichat's third article in the 1798 *Mémoires* presented a theory which he made the

[13] Ibid., pp. 193–194.

[14] Xavier Bichat, 'Description d'un nouveau trépan', *Mémoires de la Société Médicale d'Emulation*, 1798, **2**: 277–282; 'Mémoire sur la fracture de l'extrémité scapulaire de la clavicule', ibid., 309–322; 'Description d'un procédé nouveau pour la ligature des polypes', ibid., 333–338.

[15] Bichat, 'Mémoire sur la membrane synoviale des articulations', ibid., 351–370.

[16] Bichat, 'Dissertation sur les membranes, et sur leurs rapports généraux d'organisation', ibid., 371–385.

basis for all his subsequent assumptions concerning vital functions. He had discovered, he asserted, that animal nature breaks down into two essentially distinct "lives". The more basic or organic life, which is common to both the plant and animal kingdoms, functions primarily to nourish an animal and to maintain its existence apart from the inert and inorganic world that surrounds it. With its higher or animal life, a creature becomes conscious and develops a relationship with objects external to it.[17] This notion, in turn, was incorporated as a major theme in one of Bichat's books, the *Recherches physiologiques sur la vie et la mort*, which was ready for sale in May 1800.

In *La vie et la mort*, Bichat elaborated his theory on the nature of living matter. He opened the work with his now famous definition of life as "the totality of those functions which resist death". However inadequate that may seem to be from the viewpoint of the modern philosopher of science, the meaning of those words seemed clear enough to most of Bichat's readers at the turn of the century. At the root of Bichat's vitalist notions was the assumption that nature is in every respect divided into two great realms, living bodies being subject to laws and principles different from those of the physical realm. Bichat presented his investigation of the former realm in *La vie et la mort*. He stated what he considered some of those laws and principles to be. At the same time, he determined that living matter possesses five vital properties which derive from the animal's ability to perceive and to move, from its sensibility and contractility.

Bichat wrote modestly to his parents that his new work had met with some success.[18] Such evidence as there is suggests that he was already recognized as an eminent member of the medical community even before he had any sort of official position. A Leiden physician named Sandifort paid him a considerable compliment when he wrote to Hallé in Paris that "In six years, your Bichat will have surpassed our Boerhaave."[19] There were, however, the inevitable detractors also. A young professor of anatomy and physiology who did work similar to Bichat's and who might have been his rival as the rising star of Paris medicine once accused Bichat of being "a vile plagiarist".[20] Possibly with some justification, Richerand claimed priority for himself and for others for some of the ideas expressed in *La vie et la mort*.[21] As we shall see later, a great many of Bichat's ideas were borrowed from a variety of medical men. No one works in isolation, of course, but part of the problem was that Bichat rarely acknowledged his intellectual debts. This would not, in fact, be the only occasion on which such a charge would be levelled at him, though not again with such vehemence. Bichat's failing was common enough in the eighteenth century. In any case, he freely

[17] Bichat, 'Mémoire sur les rapports qui existent entre les organes à forme symmétrique et ceux à forme irrégulière', ibid, 477–487.

[18] Quoted by Genty (1933), op. cit., note 1 above.

[19] Buisson, op. cit., note 1 above.

[20] Anthelme Richerand, 'Réflexions critiques sur un ouvrage ayant pour titre: Traité des membranes', *Magasin encyclopédique*, 1800, **6**: 260–272. Othmar Keel, 'Les conditions de la décomposition "Analytique" de l'organisme: Haller, Hunter, Bichat', *Études Philosophiques*, 1982, no. 1, 37–62, refers to this denunciation, claiming that Richerand knew that Bichat's notion of membranes was lifted without acknowledgement from one A. Bonn.

[21] Anthelme Richerand, 'Essai sur la connexion de la vie avec la circulation', *Décade philosophique*, 29 June 1800.

incorporated a multitude of ideas which were floating about at the time concerning the nature of life and its functions and he combined them into a unique and coherent system of physiology. He believed that in so doing he was creating an innovative and sound theory into which all future observations and experiments made by medical scientists would be integrated. Many people were subsequently to agree with his own evaluation of his work.

In the winter of 1799, Bichat had to obtain corpses from the guillotine to supplement an inadequate supply. To do certain experiments on the subject of violent death, he arranged to have access to bodies immediately after their execution. The grisly observations he made on them provided him with the material for the second part of *La vie et la mort*.

That same winter, Bichat stopped attending the meetings of most of his medical and scientific societies, pleading overwork. The government, nevertheless, expected him to continue to do his patriotic duty. The municipal authorities named him to examine conscripted citizens who asked for exemptions from army service for reasons of health. For France, this was a period of protracted military involvement throughout Europe and in Egypt, and it seems that there were plenty of draft dodgers. Bichat was also bothered by pressure from his parents to return home to Poncin.[22] He did not yield to it, and, in fact, he never saw his parents again after he left Poncin for Paris in 1794.

Meanwhile, Bichat did not yet possess a single official title or degree. In a letter to the Minister of the Interior, he requested a position as *médecin surnuméraire* at the Hôtel-Dieu. The administrative commission of hospitals, however, awarded him the lower rank of *médecin expectant* at that institution in January 1801. Though it was less than he had hoped for, it gave Bichat the opportunity to visit patients in hospital, one of his favourite occupations, according to the eulogy of him which was delivered before the Société Médicale d'Émulation.[23] In February, Bichat offered himself as a candidate for a recently vacated chair of anatomy and physiology at the Medical School. That position went to a physician named Duméril, who was already at the school, and Bichat had to content himself with private teaching and experimentation.[24]

Bichat's major work was a four-volume *Anatomie générale appliquée à la physiologie et à la médecine*, which went on sale in August 1801. In it, one finds all the assumptions of his vitalist thesis serving as a substructure for an elaborate anatomical analysis of living matter. Bichat presented here his tissue theory of anatomy to the medical world. In many respects, it was an extension of the *Traité des membranes*, for in it, he lists the membranes previously examined among the twenty-one elements of the body. Though Bichat had borrowed many notions from his fellow physicians in order to produce this work, the tissue theory as a whole was unique and innovative,

[22] A letter from Bichat to his parents is quoted by Kervella in op cit., note 1 above, pp. 50–53. In it he was pleading for understanding from his parents. He mailed them a copy of the *Traité des membranes* at the same time; he had dedicated the work to his father.

[23] A. F. T. Levacher de la Feutrie, 'Eloge de Marie-François-Xavier Bichat', *Mémoires de la Société Médicale d'Emulation*, 1803, **5**: xxvii–lxiv. This eulogy is fairly typical of its time in so far as it abounds in flattery but is deficient in details and analysis.

[24] The letters are quoted by Genty (1934–35), op. cit., note 1 above, pp. 181–182.

although it was seen by many persons, with considerable justification, to be a kind of echo in the life sciences of Antoine Lavoisier's work in chemistry two decades earlier. Lavoisier had composed a list of the material elements which make up the compounds that preoccupy the chemist and the physicist. The great variety of these compounds, he claimed, is merely the consequence of the proportions and arrangement of these elements.

Bichat examined the unique physical and vital properties of each tissue. The details of that work were altered significantly by Bichat's successors in the next century. Its vitalist underpinnings were discarded by most physiologists within a few decades. Nevertheless, the concept basic to the tissue theory has always been conceded to be sound, thereby ensuring that Bichat has retained a place of some importance in medical history.

The arguments which Bichat presented on behalf of the tissue theory were based upon a remarkably large number of experiments and observations. In these busy years, Bichat was assisted by his cousin Buisson and by a lively young student named Philibert-Joseph Roux. Roux joined Bichat's course at the anatomical theatre when it first opened in 1798. Bichat took a liking to the bright young man, who remained with him for the last four years of Bichat's life. Though Buisson was a conscientious worker, it seems that he was too pious and even puritanical to be an attractive or comfortable companion. Roux, on the other hand, succeeded in interesting his ascetic young mentor in the theatre, which they attended night after night. This diversion for the formerly single-minded Bichat was a bit of a problem for his publisher, who was driven to charge a young apprentice named Chaudé with the task of claiming the manuscript of *Anatomie génèrale* page by page. Chaudé reported that Bichat lived amidst his anatomical specimens and their debris to such an extent that it was difficult to separate the various bits on his table into lunch and experiment. When pressured for text, he used merely to compose parts of his book on whatever bits of paper fell to hand.[25] It emerged, as we shall see later, with major inconsistencies in it, perhaps as a result of this pressure.

Buisson and Roux completed Bichat's final published work after his death. The *Traité d'anatomie descriptive* was a synthesis of his vitalist theories and of his tissue work. In the preliminary discourse to the book, Bichat related that the object of the *Anatomie génèrale* had been to decompose the body into its parts, whereas that of the *Anatomie descriptive* was to reverse the process so as to discuss the combinations of those parts into organs, systems, and functions. In this five-volume work he set out to examine the body system by system, studying the apparatus of locomotion, voice, sensation, digestion, respiration, and so on. Death interrupted the work when he was in the middle of the third volume. Buisson completed it and composed volume four, while Roux did the final one.

In his last winter, Bichat taught a course in pathological anatomy at the Hôtel-Dieu. We have a record of his lectures in the form of notes taken by one of his students. They demonstrate quite clearly that Bichat's object was to extend the

[25] This is related by Blanchard, op. cit., note 1 above, pp. 16–25. Blanchard heard the story from an aged hospital surgeon, who had been a friend of Chaudé many years before.

principles of his vital theory and his notion of tissues to include the study of pathology.[26] That winter, Bichat continued to do dissections, to practise vivisection, and to teach about eighty students. Genty lists many student dissertations that took their point of departure from the therapeutic principles of Bichat.[27] According to Buisson, he opened more than 600 cadavers in his last year. He had the temerity to work amidst odours that drove away even his most dedicated students. Just before he died, Bichat was considering a new edition of *La vie et la mort*; he had begun a series of experiments which grew out of *La vie et la mort* and had to do with the chemical phenomena of respiration; with a colleague he was performing experiments on galvanism. He began to teach a formal course in materia medica in May 1802. He reportedly sought for the basis of pharmacological action of drugs in their effect on the five vital properties, assuming that illness was the consequence of the deviation of these properties from some normal state.

Bichat fainted while descending a set of stairs in the Hôtel-Dieu on 8 July 1802. The cause of his illness is not clear. He had suffered from gastric trouble and from jaundice since the beginning of 1801, and his final illness could have been a development of some problem associated with those symptoms. On the other hand, it is possible that he was infected with some contagious fever while examining skin putrefaction in a corpse. His fellow physicians were thrown back on to their traditional treatments. Bichat was given emetics and bled by the application of leeches. A merciful coma released him from the ministrations of contemporary medical practice a few days before he died on 22 June.

Bichat's death came at the end of an important era in medical history. He had lived largely in the eighteenth century, a time that Lester King has described as "the adolescence of present day medicine".[28] It was then that the science of physiology, the specific discipline which deals with the functions and processes of living organisms, was developed. In that hundred or so years, many fundamental problems important to the subsequent development of medical and life science were identified and examined. Bichat's work belongs to that time and fits squarely into its methods and assumptions. For example, all of Bichat's observations and theories were designed so as to fit his vitalist theory. It was his consistent guide in his speculation and his experimentation. Always, he treated the vital properties as the link connecting all the branches of medicine. Bichat's ideas concerning life, its separation from non-life, and its unique vital properties became part of the theoretical and intellectual stock-in-trade of physicians. His reputation for at least the first two decades of the nineteenth century were very high. It was Magendie, a member of the Paris Medical Faculty, who first discussed Bichat's vital theory with a view to demonstrating that it must be

[26] Xavier Bichat, *Anatomie pathologique*, Paris, Baillière, 1825. A recently discovered manuscript in the Bibliothèque de l'École de Médecine de Grenoble is reported to be a richer version of these lectures than Beclard's manuscript. It is discussed by Jean Monteil in 'Le cours d'anatomie pathologique de Bichat: un nouveau manuscript', *Presse méd.*, 1964, **72**: 3163–3166.

[27] A complete bibliography of Xavier Bichat has been compiled by Geneviève Nicole-Genty, 'Bibliographie', in Genty (1972), op. cit., note 1 above, pp. 296–317. This is broken down into 'Oeuvres de Bichat', 'Travaux sur Bichat', 'Manuscrits de Bichat', 'Iconographie', and a 'Chronologie de Bichat et de son époque'.

[28] Lester S. King, *The medical world of the eighteenth century*, University of Chicago Press, 1958, p. xvi.

abandoned in order that physiology and its cognate sciences be established upon physical and chemical foundations.[29] Much of Magendie's work,and especially that of his student Bernard, was designed to show that physiology is as deterministic as physics.

Bichat was persuaded that vitalist notions of the living body were correct for reasons very much like those articulated by physicians who had received their medical training at Montpellier. By the middle of the eighteenth century, most lecturers at that school had rejected iatromechanism, the popularly held notion which maintained that the body is fundamentally an intricate and elaborate machine. Bichat was familiar with their work. Many of them, as well as persons from other European medical schools, devoted a considerable amount of time and energy to studying the sensibility and contractility of living matter. The observation and analysis of an organism's sensation and motion contributed enormously to the progress of physiological theory. Bichat created his influential physiological system from an amalgam of pro-vitalist convictions and notions concerning the reactivity of the bodily parts. We shall examine a few of these major currents in eighteenth-century physiology and their influence on Bichat's work in the remainder of this book.

[29] François Magendie, *An elementary compendium of physiology*, trans. with notes by E. Milligan, Edinburgh, John Carfrae, 1823, pp. 10–14.

II

ANIMISM, VITALISM, AND THE MEDICAL UNIVERSITY OF MONTPELLIER

Bichat was one of the last medical theorists of any particular influence to insist uncompromisingly that physics and chemistry were separate sciences from physiology, making specious the application of their principles to the study of living processes. He saw the science of physiology as a unique discipline concerning itself with activities that have no counterpart in the sciences of the physical world. His statement in the *Recherches physiologiques sur la vie et la mort* that life is defined simply as "the totality of those functions which resist death",[1] reveals more about his fundamental view of the organism than might appear at first glance. It tells us essentially that he saw each living body as an organic unit besieged by the forces belonging to the surrounding inorganic world. Although living forces are in organized matter only for a limited time, they are able to exert their dominance over physical ones. He would argue that it is by means of these combative vital forces that an organism grows, reproduces, is nourished, and responds to its environment. They leave the body very gradually if death is natural or lingering, or they flee quickly if it is accidental and sudden. Thereafter, the physical forces reassert their dominance, causing the organism to decompose and gradually to become one with the simpler, more stable and predictable realm of inorganic nature.

The conviction that life sciences are unique and to be distinguished from inorganic ones was widely shared by many of Bichat's contemporaries. It put him into a vitalist tradition that can be traced back at least to Aristotle. Eighteenth-century vitalists contributed a great deal to the development of the medical sciences and especially to their philosophy. In this chapter, I examine the evolution of certain vitalist notions, many of them developed by French physicians in the post-Newtonian period.

In 1700, the orthodox European medical man was a mechanist and a dualist, who believed that the living being is a composite of a material body and an immaterial soul. That soul, he assumed, oversees the voluntary, willed and rationally determined functions of the body. The remaining activity of that same body is automatic and involuntary. It occurs mechanically, for an organism, being part of nature, is basically no more than an intricate network of pulleys, pipes, and conduits whose activity is in accordance with mechanical and hydraulic principles.[2] This view was soon challenged on the grounds that mere mechanism is not adequate to account for the complexity of living phenomena.

By 1800, what is known in broad terms as the mechanist-vitalist debate was generally conceded to have been resolved in favour of the latter position. Indeed, Bichat's particular viewpoint was largely acceptable to most of his colleagues and con-

[1] Xavier Bichat, *Recherches physiologiques sur la vie et la mort*, Paris, Brosson, Gabon, 1801, p. 1.

[2] Two recent general works on medical philosophy in the enlightenment which touch on the topics I discuss here are Lester S. King, *The philosophy of medicine*, Cambridge, Mass., Harvard University Press, 1978; and François Duchesneau, *La physiologie des lumières: empiricisme, modèles et théories*, The Hague, Nijhoff, 1982.

temporaries. Among the physicians and physiologists in the latter part of the eighteenth century who considered themselves to be anti-mechanists, there may still have been some animists, so-called because they considered all living activity to be explained by the control of the body by a rational soul or *anima*. They shared a dualist theory of nature with the mechanists, for both believed that matter is inert substance and hence capable of motion only if subjected to an external force. Other vitalists postulated some specific life-conferring principle which coexists with the body and the soul as a kind of third substance overseeing the automatic and unconscious acts of life. Most other vitalists, especially after the mid-eighteenth century, were monists who believed that vital properties such as sensibility and contractility reside in the bodily parts themselves. Although they considered these properties to be distinct from and irreducible to the physical ones, they thought them to be intrinsic in the material of the body and a product of its organization.

Bichat and other monists are frequently labelled "organicists" or exponents of a theory of "organicism". Indeed, the term confers more precise information than the more general "vitalists", for it ensures that its adherents will not be confused with animists. Organicism was, nevertheless, a sophisticated form of vitalist theory, the end-product of a long debate concerning the nature of living matter. Many organicists whose views I shall consider admitted that they adjudged themselves to be intellectually and philosophically relatives of the animists, united to them in their mutual rejection of mechanism. I believe, therefore, that the more familiar vitalist label, albeit imprecise, is useful as long as we remember that vitalists were far from being a homogeneous group.

It should be clear by the time that we come to examine Bichat's vitalist theory in detail that he assimilated his fellow physicians' ideas with considerable ease. His largely synthetic work marries prevailing notions of the late eighteenth century to his own observations and assumptions. He was strongly influenced by the ideas of the Montpellier physicians, to whom is owed much of the credit for at least the earliest development of eighteenth-century vitalist theory. It all began at that ancient and venerable institution in 1737, when a young lecturer, François Boissier de Sauvages, introduced animist assumptions into his teaching, provoking outraged protest among his conservative and stolidly iatromechanist colleagues. In time, however, his viewpoint commended itself to growing numbers of faculty and especially to students. Although most of them did not long remain animists, the language of anti-mechanism was conceptually liberating.

The basic positions in the mechanist-vitalist dispute had been effectively drawn up by the mid-seventeenth century. Those persons who subsequently engaged in the debate about the nature of life and its activity owed their fundamental assumptions to the work of one of two men. Both mechanists and animists were heirs to the dualist and mechanical philosophy of *René Descartes* (1596–1650), while the organicists relied considerably on the work of *Jan Baptista van Helmont* (1577–1644). Both men felt themselves to be rebels against the stultifying remnants of the Aristotelian and Galenic notions which persisted in the schools they attended in their youth. Their respective philosophical speculations, however, produced vastly different results.

Born into an upper-class family in Brussels, Helmont received a degree from the

University of Louvain, where he came into contact with the literature of the hermetic philosophy replete with mystical, alchemical, and astrological elements. Its language and imagery coloured his adult work. His writing, which addressed chemical, medical, and philosophical themes, was collected and published as the *Ortus medicinae* after his death. In 1661, it was translated into English and published under the title *Oriatrike or physick refined.*

Helmont claimed a profound disillusion with his youthful studies. In 1594, he wrote: "It dawned upon me that I knew nothing and that what I knew was worthless." Longing for truth and science, he recalled years later, he set out on a quest for certainty which led him away from the traditionalists and the schoolmen. A deeply religious man, he turned to God to show him a way out of his disillusion. "I prayed to the Prince of Life for the stamina to contemplate the naked truth and love it *per se.*" The work of Galen, Hippocrates, Avicenna, and more recently of Fuchs and Fernel, he claimed, left him unenlightened. "I re-read the collection of my notes and recognized my poverty, and the labours and years I had consumed angered me."[3]

About twenty years after Helmont's youthful rebellion, Descartes was similarly repelled by the shallowness and uncertainty of what he had been taught. Both men wanted to know wherein lay certainty and both rejected mere sensory evidence as misleading on the grounds that genuine knowledge is too elusive and confusing to be discovered merely with our five senses.

Helmont's enlightenment concerning the proper method to be used to acquire knowledge came to him as he sat beside his alchemical furnaces; Descartes' flash of insight occurred in a heated room in Germany. Descartes looked to clear and careful reasoning to show him how to distinguish what is true from that which merely appears to be so. He proceeded from the one truth which appeared to him to be irrefutable – "I think, therefore I am."[4] Helmont, on the other hand, claimed to have had a vision of his own indivisible and immortal soul receiving the light of knowledge from God. Believing that just such flashes of God-given insight are the basis of all knowledge, he sought intellectual and spiritual enlightenment through prayer and meditation. Accordingly, he eschewed the quest for nature's secrets by the application of reason, claiming that "our minds ought to be intellectual but not rational".[5]

The result of Descartes' fireside meditations was his dualist view of nature, a philosophy that divided the world into two distinct realms, matter and the immaterial. Descartes distinguished the material body from the mind, from life, and from its soul. According to him, the living body is properly understood as an intricate machine, a

[3] Jan Baptista van Helmont, 'Studia authoris', *Ortus medicinae*, Amsterdam, L. Elzeverius, 1648, 16–19. Translated and quoted by Walter Pagel, 'The reaction to Aristotle in seventeenth-century biological thought', in Edgar Ashworth Underwood (editor), *Science, medicine, and history*, 2 vols., Oxford University Press, 1953, vol. 1, pp. 491–492.

[4] René Descartes, *Discours de la méthode* in *Oeuvres de Descartes*, ed. by Charles Adam and Paul Tannery, 11 vols., Paris, Vrin, 1973, vol. 6, pp. 31–40. Descartes' "nonvitalistic, material-mechanical" ideas are discussed by Thomas S. Hall, *Ideas of life and matter*, 2 vols., University of Chicago Press, 1969, vol. 1, pp. 250–263.

[5] Helmont, 'Venatio Scientiarum', *Ortus medicinae*, op. cit., note 3 above, pp. 24–32. For convenience, I have quoted throughout from the seventeenth-century translation of the original text. The translated reference is 'The hunting or searching out of sciences', *Oriatrike or physick refined*, London, L. Loyd, 1662, p. 15.

kind of clockwork, subject to the will and to thought which belong to the soul. This soul keeps the clockwork running. The source of its sensation and motion, it acts upon the body by means of a very fine intermediate sort of material known as animal spirits, which flow through the nervous system. Matter by itself is inert and incapable of self-motion. Descartes discussed this notion of physiological dualism and its consequences in his *Traité sur les passions et sur l'âme* of 1649, and in *L'homme*, which was published posthumously in 1664.

Helmont did not so divide matter. To him, life is immanent in bodily substance rather than being externally conferred upon it. That is, Helmont was a monist. Although he believed in the existence of a soul that is both a spiritual and rational entity, he did not consider it to be an exclusive mover of the body. Far from being a machine, the body is, he believed, an entity having its source of motion within itself.[6] Many of the most significant physiological debates of the next two centuries had to do with the conflicting viewpoints which derived from the adoption of either the monist or dualist view of matter.

Helmont's words are confusing largely because his imagery is free-wheeling, archaic, and by now, utterly unfamiliar. This is exacerbated by his rambling and incoherent writing style, which is equally difficult to follow in the original Latin and in its English translation. Until Walter Pagel produced his careful and painstaking analysis of Helmont's writings, his rambling thoughts were little understood, almost as obscure and mysterious as the faculties which he attributed to nature.[7] Reading Helmont's writings, we confront a pervasive sense of life's mystery and of its complex inter-relationship with a universe throbbing with life and spiritual forces. He conferred upon the body a formidable array of chemical and spiritual forces whose essence is never accessible to mere scientific investigation.

By means of perhaps the most famous of all experiments, Helmont determined that water alone accounted for the weight gained by a tree during growth. He concluded that all matter may be reduced ultimately to water. Spiritual and immaterial objects, on the other hand, derive from a fundamental air. The transformation of water into a tree is due to activity of an in-dwelling seed which shares the nature of both spirit and matter, that is, of both air and water. Such seeds exist in all natural objects and they contain information which determines the future nature or form not only of an adult plant or animal, but of rock or metal as well. "Whatever therefore cometh into the

[6] Pagel, op. cit., note 3 above, pp. 489–509.

[7] Probably the single best source for understanding Helmont's complex theory is Walter Pagel's 'The religious and philosophical aspects of Helmont's science and medicine', *Bull. Hist. Med.*, 1944, Supplement No. 2. In this work, Pagel argues that Helmont can properly be considered to be the first biochemist, since, with Paracelsus, he was one of the first to express vital phenomena in chemical terms as well as one of the first to use such instruments as the balance and thermometer in medicine. Pagel elucidates some of the complex and confusing terms that occur in Helmont's mystically inclined works. He demonstrates that Helmont's particular viewpoint owes much to his religious zeal. See also, Walter Pagel, *Joan Baptista Van Helmont: reformer of science and medicine*, Cambridge University Press, 1982. Other general sources are Lester S. King, *The road to medical enlightenment, 1650–1695*, London, MacDonald, 1970, pp. 37–62; King, op. cit., note 2 above, pp. 34–40, 125–134; Jacques Roger, *Les sciences de la vie dans la pensée française du XVIIIe siècle*, Paris, Colin, 1963, pp. 98–103; and J. Mepham, 'Johann Van Helmont, 1574–1644', in Rom Harré (editor), *Early eighteenth-century scientists*, Oxford, Pergamon, 1965, pp. 129–157.

World by Nature, it must needs have the Beginning of its motions, the stirrer up, and inward director of generation."

In each seed there is a dormant *archeus*, a kind of sensitive soul which exercises a dynamic control of the body and provides its special character. "The Governor of generation", the *archeus* is a spiritual principle which transforms matter during growth and development. In the rendering of the seventeenth-century translator, "in things soulified, he walketh thorow all the Dens and retiring places of his Seed, and begins to transform the matter, according to the perfect act of his own Image. For here he placeth the heart, but there he appoints the brain, and he everywhere limiteth an unmoveable chief dweller, out of his whole Monarchy, according to the bounds of requirance of the parts, and of appointments."[8]

Helmont believed that, like the heavenly sun, the bodily "principle of life", is centrally located, radiating its influence in all directions. Accordingly he said it resides in the upper part of the stomach[9] where the mysterious process of transformation of food into flesh and blood begins.[10] A kind of organic alchemist, the *archeus* presides over bodily processes by controlling many subsidiary *archei* residing in the organs and smallest bodily parts. They account even for illness. Infection, wrote Helmont, is due to the invasion of the body by foreign *archei* which alter and disrupt normal physiological processes.

Helmont considered the *archeus* to have come into being only with the fall of man from grace in the Garden of Eden. In addition to this mortal soul, man has an immortal one, the *mens*, centred in the brain. The two were wedded together until death became part of the human experience due to sin. Now the *mens* is tied to the body only as long as life is present.[11] Although he tried to free himself from peripatetic imagery, we cannot help but be reminded here of the classical triumvirate of souls invoked to explain human activity and to distinguish different categories of life. So, like the Greeks, Helmont assigned consciousness and the intellect to the government of an immortal soul which acts on the body through a vital principle. A product of digestion, it rises to the brain to govern the senses and mental functions in general.[12]

Any attempt to organize or to summarize Helmont's thoughts inevitably produces some distortion. His words are a labyrinth of *blas*, ferments, gases, *archei*, seeds, and souls, all of whose actions intricately intertwine to control the processes of life and to link it to the spiritual realm. Still, persons like Pagel who have thoroughly studied his

[8] Helmont, 'Archeus faber', *Ortus medicinae*, op. cit., note 3 above, pp. 40–41; 'The chief or master-workman', *Oriatrike*, op. cit., note 5 above, pp. 35–36.

[9] Helmont, 'Sedes animae', *Ortus medicinae*, op. cit., note 3 above, pp. 288–293; 'The seat of the soul', *Oriatrike*, op. cit., note 5 above, pp. 283–288.

[10] One of the most important works on this subject is 'Sextuplex digestio alimenti humani', *Ortus medicinae*, op. cit., note 3 above, pp. 208–225; 'A six-fold digestion of human nourishment', *Oriatrike*, op. cit., note 5 above, pp. 205–221. In this work the author examined what he believed to be a six-stage process of the transmutation of ingested food into bodily flesh. The first occurs in the stomach under the supervision of the spleen, which supplies a ferment to transform food into a cream. The second is in the duodenum, where bile neutralizes acidity; the third begins the process of making blood in the liver. And so on, until we reach the sixth stage of digestion, which occurs in "particular Kitchins of the Members", or in the tissues themselves. Each step is overseen by a "ferment", which is perhaps very roughly comparable to a chemical catalyst.

[11] Helmont, 'Sedes animae', op. cit., note 9 above.

[12] Helmont, 'Spiritus vitae', *Ortus medicinae*, op. cit., note 3 above, pp. 195–201; 'The spirit of life', *Oriatrike*, op. cit., note 5 above, pp. 192–197.

work see much there that was formative for the subsequent development of physiological theory. Among the physicians who were to express admiration for Helmont were many from Montpellier, including Théophile de Bordeu, a most important figure in the development of eighteenth-century vitalist theory. In our own time, Helmont has even received credit for being the founder of biochemistry because of his particular iatrochemical orientation and use of measuring instruments.[13]

Although Helmont was a creative and formative thinker, a great gulf separates him from his eighteenth-century successors. Between the time of his death in 1644 and the end of the century, the study of nature developed more profoundly and dramatically than at any other time in history. The method and the language of science as we know it largely took shape in those years with the emergence of the mechanical philosophy. Helmont's mystical and spiritual imagery and his visionary approach to the study of the organism became inadmissible. Furthermore, the separation of body and soul, of matter and spirit, which Descartes had posed for natural philosophy, thoroughly seared itself into the consciousness of European scientists and philosophers so that none of Helmont's successors could mingle the physical and spiritual realms as blithely as he had done.

Although the most striking achievements of the scientific revolution, which so profoundly altered man's conception of the natural world, was largely in physics and astronomy, the study of living nature was affected also. Perceiving the quantitatively based new physics to be remarkably predictable and coherent, many physicians and natural philosophers made it their goal to apply its assumptions and methods to their particular investigations. The most famous achievement in seventeenth-century medical science was the demonstration of the circulation of the blood. It is instructive to note that in his *Discourse on method*, Descartes took it to be supportive of the philosophy of biological mechanism even though Harvey himself was to emerge as a confirmed vitalist and even an Aristotelian. By the 1660s, there was emerging in Europe an iatromechanical school whose philosophy grew out of an intricate intertwining of physics and physiology. It was the epistemological and conceptual offspring of the mechanical philosophy. There was thereafter an increasing tendency to ask questions about the organism which invited answers based on mechanical, hydraulic, or corpuscularian assumptions. It was taken as axiomatic that in and of itself, matter is inert substance. The study of all natural phenomena was reduced to the examination of the motion of matter under the influence of forces external to itself. The greatest triumph of such an approach was, of course, Isaac Newton's demonstration of the laws of attraction in the *Principia mathematica.* Even before 1687, however, iatromechanism was on its way to becoming the theoretical substructure upon which physicians and philosophers proceeded to construct a new organic theory. The approach, for example, of Helmont came to be discarded as meaningless. On the other hand, the mechanical philosophy of Descartes, Galileo, Gassendi, and a great many other natural philosophers spoke eloquently to contemporary needs and assumptions.

Iatromechanical works first appeared in large numbers in Italy among persons who

[13] Pagel (1944), op. cit., note 7 above.

had been schooled in Galileo's mechanics. Physicians engaged in dialogue with mathematicians and students of mechanics, thereby producing a fruitful new view of organic function. Mathematics and physics were brought powerfully to bear on the study of human anatomy and physiology, and the image of the body as an intricate hydraulic machine began to preoccupy the students of living functions. Giovanni Borelli, for example, who held a post as lecturer in mathematics in Pisa in the 1650s, acquired an enthusiasm for medical questions from contact with the physicians Marcello Malpighi and Lorenzo Bellini. Probably because of this contact, Malpighi, in 1661, produced a strongly iatromechanist examination of lung anatomy and function, which he entitled *Epistola de pulmonibus* and dedicated to Borelli. In 1662, Bellini published *Exercitatio anatomica de structura et usu renum*, in which he tried to account for renal function entirely without recourse to such concepts as innate and attractive faculties. He wished to understand the parts as a mixture of solids and fluids with such quantifiable attributes as velocity, viscosity, and momentum. Another work by Bellini, his *Opuscula aliquot* of 1695, was actually organized into postulates, theories, and corollaries in obvious imitation of mathematics. Borelli's *De motu animalium* of 1680 is merely the most famous of a considerable number of works that treated the subject of the mechanics of muscular motion. In general, many works in that period appeared in which muscles, glands, the nervous system, and so on were analysed with the help of geometric language and presented in a mechanical and mathematical format. The Fellows of the Royal Society of Great Britain, as well as medical people in all of Europe, were seduced by the innovative new programme so that by the end of the seventeenth century, virtually every individual even minimally conversant with the new science had to give serious thought to the mechanical theory of living matter.[14] The first substantial challenge to that position came early in the next century in the form of the animist philosophy.

The animist reaction occurred largely because the iatromechanical hypothesis left its proponents with the problem of having to explain what moves the living machine, at a time when it was taken as axiomatic that matter is passive substance with no innate ability to generate motion. Willed and conscious activity such as that of voluntary muscles posed no particular problem. As most persons had always done, mechanists simply ascribed it to a soul or rational and even spiritual principle supposed to reside in the brain. So, in fact, did the animists and many vitalists. By so doing, they were accepting a long-standing albeit ambiguous definition of the soul. The *psyche* of ancient Greek usage was at once a spiritual, a vital, and a rational principle. Aristotle and Galen, for example, used the word to denote the source of mind, of consciousness and of life itself. As such it was a vital force or principle. In classical Latin, a distinction was made between the *anima*, which is a spiritual or vital principle, and the *animus*, which is a rational one. Europeans subsequently tended to conflate the terms such that the early translators of the Bible, for example, wrote only of the *anima*. Most modern European languages have only a single word to denote

[14] For a general discussion of mechanism, see Roger, op. cit., note 7 above, pp. 207–249; and King, op. cit., note 2 above, pp. 95–124. Two approaches to iatromechanism, the Galilean and the Cartesian, are discussed by François Duchesneau, 'Malpighi, Descartes, and the epistemological problems of iatromechanism', in M. L. Righini Bonelli and William R. Shea (editors), *Reason, experiment and mysticism in the scientific revolution*, New York, Watson, 1975, pp. 111–130.

soul. Similarly, in the eighteenth century, one word retained both its spiritual and rational connotations. Hence for physicians especially, the soul had a kind of dual role in the body.[15]

Most of the functions of the body, of course, are involuntary, automatic and unperceived. Much of the discussion has to do with the question of their origin. What is the source of power behind the heart, the digestive system, nourishment, and growth? We have seen how Helmont's *archeus* served to explain these functions. The mechanists deemed it sufficient to assume that the body is a massively complex automaton. Once set in motion, its parts interact to produce an involved series of motions which succeed one another of necessity. To use Descartes' own imagery, a living body is like an infinitely elaborate clockwork whose parts act automatically. Like a clock, it slows down and finally stops if left to itself. But that sort of explanation was altogether successful only so long as one posed only particular questions, and those in a particular way.

Borelli provides us with an example of the dilemma in which iatromechanists sometimes found themselves. In *De motu animalium*, he had difficulty in accounting for the action of the heart in purely mechanical terms. It is an involuntary organ and as such, it ought to be independent of the soul. Indeed, an excised heart is frequently observed to beat for a time. On the other hand, mental or emotional states are often perceived to affect its motion. This apparently conflicting evidence drove Borelli to speculate that, in the case of the heart, the soul is acting unconsciously.[16] The identification of a growing number of such inconsistencies in the mechanists' accounts provided fertile ground for an anti-mechanist reaction.

The mechanist-animist debate largely took shape in the first four decades of the eighteenth century. It can be illustrated by a glance at the notions of three men who, in those years, were perhaps the most famous physicians in Europe. They are *Friedrich Hoffmann* (1660–1742), a Cartesian; *Herman Boerhaave* (1668–1738), a mechanist; and *Georg Ernst Stahl* (1659–1734), an animist. The problems to which they addressed themselves and the divergent hypotheses which they advanced to deal with them laid the basis for the speculations and the experiments of many enlightenment physicians and philosophers, including the Montpellier physicians and Bichat.

Hoffmann completed his medical training in 1684. A year later, while travelling in Europe where he met the English chemist Robert Boyle, he received an invitation from Frederick III, the Elector of Brandenburg and the future King Frederick I of Prussia, to become the Professor of Medicine at a new university just established in his native Halle. Some nine years later, Hoffmann, in turn, invited Stahl to join him in

[15] Theodore Blondin, 'Du vitalisme animique. Études générales devant servir d'introduction à la physiologie de G. E. Stahl', *Oeuvres médico-philosophiques et pratiques*, 6 vols., ed.by Theodore Blondin, Paris, Baillière, 1860, vol. 3. Blondin demonstrates the trap awaiting the unwary historian who does not know of the confusion surrounding the definition of the word *soul*. Blondin particularly approved of Stahl's animism, arguing that he had struck a blow on behalf of God and religion in a world losing its faith. He rhapsodized about Stahl's piety and his virtuous attack on atheism and materialism, although he confused theology and physiology by so doing. See also, Roger K. French, *Robert Whytt, the soul, and medicine*, London, Wellcome Institute, 1969, pp. 117–148; and L. J. Rather, 'G. E. Stahl's psychological physiology', *Bull. Hist. Med.*, 1961, **53**: 37–49.

[16] This example comes from Roger K. French, 'Sauvages, Whytt and the motion of the heart: aspects of eighteenth-century animism', *Clio medica*, 1972, **7**: 35–54, see 37–38.

Halle. But long before Stahl wrote anything on the subject of the living organism, Hoffmann was speculating in print on the nature of the living body.

His *Fundamenta medicinae* of 1695 was a textbook of general medical theory, ranging over physiology, pathology, symptomatology, hygiene, and therapeutics and touching upon the methodology appropriate to each. It was intended to provide its readers, many of whom were medical students, with a general viewpoint on medicine. That viewpoint turned out to be iatromechanism wedded to many of Hoffmann's Cartesian assumptions about mind and matter.[17] In the introduction, he affirmed a wish that medicine should become a science employing a "simple, clear mathematical method". Nature, he wrote, is mechanical. Matter and motion are its first principles. "Changes and alterations of the entire universe are due to motion." Accordingly, God's role in His creation is that of a supreme mechanic, the "prime mover of all things [who] . . . produces things by statics and mechanics and maintains everything in its equilibrium."[18]

It seems to follow from such a viewpoint that "Medicine is the art of properly utilizing physico-mechanical principles in order to conserve the health of man or to restore it if lost"; "Like all of nature, medicine must be mechanical . . . . The first principles of mechanics are matter and motion"; and finally, "Life is achieved by causes which are wholly mechanical. The mind does not bring life to the body, nor is life oriented to the mind, but rather to the body."[19] So it is, for example, that the production of bodily humours is explained solely by filtration. "The viscera do not separate out the excremental humours through any ferment that precipitates the humours to be secreted, but all secretion of the bile, urine, sweat, phlegm, saliva, is accomplished by mechanical means and a particular manner of filtration."[20]

Since the passivity of matter was a fundamental tenet of the mechanical philosophy, Hoffmann had to address the question of how the material parts are to be activated. He assumed that some spirit or prime mover must exist to move the machine. While life is present, he wrote, there is a continuous movement of the fluid parts of the body through the solid ones. These fluids, including the animal spirits, the blood, and the lymph, excite the solid parts inducing them to perform appropriate movements. Health exists when the motion is well ordered. Death is simply the destruction of motion. With most of his contemporaries, Hoffmann believed that the initial source of the motion is the soul or *anima* which possesses both material and spiritual properties. His own words say it best:

> In our machine, the first principle of motion is the soul which you may, if you want, designate as nature, or spirit endowed with mechanical powers, or a most subtle ethereal matter acting in an ordered and specific fashion . . . .
>
> The first cause of the powers existing in the soul [*anima*] is God himself, on whom all things depend,

[17] The work is easily accessible since Lester S. King's English translation of it as Friedrich Hoffmann, *Fundamenta medicinae*, London, MacDonald, 1971. King summarized Hoffmann's viewpoint in an 'Introduction', pp. iv–xxv. See also, King, op. cit., note 2 above, pp. 34–40. A more ambitious analysis of Hoffmann's mechanist system is presented by Duchesneau, op. cit., note 2 above, pp. 32–64.

[18] Hoffmann, 'To the reader', op. cit., note 17 above, pp. 1–4.

[19] Ibid., pp. 5–11.

[20] Ibid., pp. 28–34. This quotation is interesting for our purposes because some fifty years later, Théophile de Bordeu would deal a decisive blow to the iatromechanist viewpoint by using the evidence produced by glandular activity, as we shall see later in this chapter.

including the existence and functions of our soul.

The animal spirits are not the vital principle or soul itself but the soul uses them as instruments for achieving its functions.

The functions of the soul are to move the spirits and to direct them to produce particular motions.[21]

In addition to its mechanical capacities, the human soul is endowed with the power of thought and reason. It is an "immortal substance stemming from the decree of God himself".[22]

Animal spirits exist largely in the nerves. Too fine to be seen, they are the will's instrument for achieving sensation and muscular motion. In response to the soul's direction, they move in accordance with mechanical principles and physical laws. Other words written by Hoffmann, however, concede that the animal spirits seem to possess some capacity for self-motion. "The animal spirits", he wrote, "have a power impressed by God, not only of moving themselves mechanically, but doing so by choice, purposefully and toward a definite goal. This power is called the sensitive soul, and it exists entirely in the most subtle fluid of the brain." Focused in what he called a "common sensory" in which external impressions are gathered, the "animal spirits are directed and pass out to all the parts which would be animated by sensation and motion".[23]

Hoffmann's work is an example of the sort of dilemma which iatromechanism, rooted in assumptions about inert matter, created for its adherents. The body required a non-material source of motion. When they considered how pure spirit is able to impose its will upon pure matter, they usually imagined some intermediate substance combining the properties of both spirit and matter. By so doing, they were according the capacity of self-motion to at least *some* matter, so as to solve the difficulties posed by the belief that it is inert.

Because Hoffmann considered himself to be a Cartesian mechanist, we treat him as such. It is true, however, that our perception of his notions can deviate from Hoffmann's own. It seems clear to me, for instance, that although Hoffmann professed to disagree with the animist position, he moved perilously close to it when he cast the soul into the role of prime mover. How close can be seen by a comparison with Stahl, who put his provocative animist notions of physiology into print some ten years after the *Fundamenta medicinae* at a time when the two men were colleagues at Halle. Stahl's fundamental tenet was that the soul is the exclusive source of all bodily motion, voluntary and involuntary alike. Often, it is difficult to say precisely where their notions differed, disagreement apparently having more to do with emphasis than with substance. It remains true, however, that in spite of the apparent correspondence between many of the two men's ideas, they considered themselves to be separated by a wide theoretical gulf.[24]

However influential the *Fundamenta medicinae* might have been initially, it was displaced in importance by the work of Boerhaave, especially by his *Institutes* which

[21] Ibid., pp. 11–13.

[22] Ibid.

[23] Ibid., pp. 23–28.

[24] For a discussion of the relationship between Hoffmann's and Stahl's ideas, see Lester S. King, 'Stahl and Hoffmann: a study in eighteenth century animism', *J. Hist. Med.*, 1964, **19**: 118–130. For a further discussion of Hoffmann's work, see King, op. cit., note 7 above, pp. 181–204.

appeared in 1707. Boerhaave's very name came to be synonymous with iatromechanism and his reputation as a teacher was unsurpassed in his lifetime. He attracted students from Europe and North America to Leiden. Because many disciples spread his fame widely, his notions governed medical instruction as far away as North America. William Cullen reported that in Edinburgh, as late as in the 1760s, he was advised not to differ from Boerhaave for fear of hurting himself or his university.

Cartesian notions had penetrated the Leiden medical school before Boerhaave received a chair there in 1703. At his inaugural address to the university, he dedicated himself to fixing the mechanist thesis upon a firm foundation of observation and experimentation. His methods must have been deemed appropriate, for his school quickly became the most celebrated in Europe.[25]

In a work discussing 'The nature and principles of physiology', we find Boerhaave's position expressed as follows:

> The solid Parts of the human Body are either membranous Pipes, or Vessels including the Fluids, or else Instruments made up of these and more solid Fibres, so formed and connected, that each of them is capable of performing a particular Action by the Structure, whenever they shall be put in Motion; we find some of them resemble *Pillars, Props, Cross-Beams, Fences, Coverings*, some like *Axes, Wedges, Leavers*, and *Pullies*; others like *Cords, Presses* or *Bellows*; and others again like *Sieves, Strainers, Pipes, Conduits* and *Receivers*; and the Faculty of performing various Motions by these Instruments is called their *Functions*; which are all performed by *Mechanical Laws*, and by them only are intelligible.[26]

How then does one account for the motion of pipes, sieves, and presses? From what primary impulse do the bodily functions proceed? Emphatic though he was that man is a composite of mind and body, Boerhaave did not speculate overmuch in his writings about the nature of the mind-body relationship, simply acknowledging it to be a "harmony established by God". The material and spiritual components of the body are such that "each has a Life, Actions and Affections differing from the Other". Nevertheless, "there is such a reciprocal Connection and Consent between the particular Thoughts and Affections of the Mind and Body, that a Change in one always produces a Change in the other and the reverse." But unconscious bodily functions depend upon the mind not at all. "There are some Actions performed by the Body without the Attention, Knowledge or Desire of the Mind, which is neither concerned therein as the Cause or Effect of these Actions."

Boerhaave may have briefly flirted with the notion that matter in an organized body might have a life or motive principle apart from the rational one. At least, that is a possible interpretation of the following words:

> The Life of the Body is, 1. To generate Motion under particular Circumstances, as the Loadstone approaches to Iron. 2. For its constituent Parts to attract each other, from whence proceeds that Resistance to the Force of external Bodies, or *Vis Inertia*. . . . In a Word, the Life of the Mind is to be conscious.

[25] Boerhaave's influential viewpoint is discussed in a number of sources including Gerrit A. Lindeboom, *Hermann Boerhaave, the man and his work*, London, Methuen, 1968; Lester S. King, *The medical world of the eighteenth century*, University of Chicago Press, 1958, pp. 59–121; Hall, op. cit., note 4 above, vol. 1, pp. 367–390; and Duchesneau, op. cit., note 2 above, pp. 103–140.

[26] Herman Boerhaave, 'The nature and principles of physiology', *Dr. Boerhaave's academical lectures on the theory of physic, being a genuine translation of his Institutes as dictated to his students at the University of Leyden*, 6 vols., London, W. Innys, 1757–73, vol. 1, pp. 80–85. The emphasis in this and other quotations is the author's own.

But unfortunately, Boerhaave dismissed ultimate questions about life and matter on the grounds that the true physician properly applies himself to restoring health. So it is that "the *primary physical Causes* . . . and the ultimate *metaphysical Causes* . . . are neither possible, useful, or necessary to be investigated by a Physician; . . . The Origin of Motion is to be look'd for in God; if we substitute any other primary Cause, we do him Injustice."[27]

As we have seen, the ease of application of mechanist or animist assumptions in the eighteenth century depended upon the nature of the questions one asked of the organism. Iatromechanists tended to be most successful when they studied the functions of specific parts. Their approach lent itself to examining, for example, muscle contraction, blood flow, and glandular filtration. It was less satisfactory when one focused upon the organism with a view to understanding its integrated and apparently purposeful activity. Although a mechanist might be convincing when he discussed the various stages of digestion, he was less persuasive when he addressed the process of nutrition. He constantly ran the risk of oversimplifying and even ignoring such complex activity as adaptation, co-ordination, and growth.[28]

Stahl addressed just such integrated and long-term processes in his writings. At a time when an anti-mechanist stance was so provocative as to border on the revolutionary, he rejected mechanism as an inadequate tool for understanding living phenomena. Stahl, who was a professor at the University of Halle from 1694 to 1715, published most of his medical writings between 1706 and 1708. The overall view of the living body that emerges from them is that of a complex, dynamic, and infinitely reactive structure.[29] It is this incessant but purposeful activity which seems to have particularly impressed him and to have directed his speculations. He believed that the reactive quality belongs to living things alone among all natural objects and it is in that reaction that he searched for clues concerning the special features which distinguish life from non-life.

In a work on the subject, *De vera diversitate corporis mixti et vivi*, Stahl remarked that the most fundamental feature of a living body is its unique complexity of form and organization. "If a mixture, an ordinary chemical composite is homogeneous and stable, a living body is naturally heterogeneous and condemned to corruption if it is abandoned to itself."[30] There is an enormous variety among living species simply because each species has its unique and complex mixture of physical elements. This very mixture is the basis of all life-related functions. It follows, therefore, that one

[27] Boerhaave, 'Of the parts and principles of physic', ibid., pp. 51–79.

[28] Discussed by Lester S. King, 'Basic concepts of eighteenth-century animism', *Am. J. Psychiat.*, 1967, **124**: 105–110; and by Hilde Hein, 'The endurance of the mechanism-vitalism controversy', *J. Hist. Biol.*, 1972, **5**: 159–188. See also, Lester S. King, 'Rationalism in early eighteenth-century medicine', *Bull. Hist. Med.*, 1963, **37**: 257–271.

[29] Stahl's medical theory is discussed in a number of works, the earliest being by Albert Lemoine, *Le vitalisme et l'animisme de Stahl*, Paris, Baillière, 1864. See also, Roger, op. cit., note 7 above, pp. 427–431; Hall, op. cit., note 4 above, vol. 1, pp. 351–366; François Duchesneau, 'G. E. Stahl: anti-mécanisme et physiologie', *Archs. int. Hist. Sci.*, 1976, **26**: 3–26; and more recently, Duchesneau, op. cit., note 2 above, pp. 1–31. Part of Duchesneau's purpose is to examine the epistemological aspect of Stahl's notion of physiology, which emerged at the time when the frontier between physiology and physics was not yet clear.

[30] Georg Ernst Stahl, *Demonstratio de mixti et vivi corporis vera diversitate*, Halle, 1738, §§X–XV, pp. 70–75. In the French translation, the reference is *Mixte et vivante* in *Oeuvres médico-philosophiques*, op. cit., note 15 above, vol. 2, pp. 366–376.

ought not to attempt to understand life apart from that mixture, and also that it is important to try to determine what sustains it.

Stahl was shocked, he wrote, to find that nowhere among the mechanists was there any attempt to arrive at a logical definition of life,[31] which he defined as "the conservation of an eminently corruptible body, the faculty or force with whose aid the body is sheltered from the act of corruption".[32] Notice that the key word is "conservation". That is, though the body is a corruptible mixture, life consists not in that mixture itself but in its preservation. Indeed, like Stahl, we observe daily that life seems to be the cause and not the consequence of organization. The breakdown of a body into its elements begins as soon as some mysterious living force or essence leaves the organism at death. Organization fails, causing the decaying parts gradually to merge into their inanimate surroundings.

Stahl believed that this conserving function belongs to the *anima*. It and it alone activates the passive matter of the body. It oversees the unwilled as well as the willed, the unconscious as well as the conscious action. Stahl avoided many of the conceptual difficulties that afflicted the mechanists by treating the distinction between conscious and automatic acts as artificial. He observed, for example, that such emotions as joy, fear, and anger affect the functions of organs which are not normally under control of the will. He did not even deny that living activity proceeds in accordance with mechanical principles. His emphasis in his writings, however, was on the soul's direction.

The soul acts on the body to conserve it, Stahl wrote, by means of motion. Fluids appear to mediate between the solid body and the immaterial mind:

> It is by motion that the *human soul* accomplishes its work *in* and *on* the body, as powerfully and as long as it can, but one cannot say in an absolute and true manner that *motion is life*. It is again by the *circulatory motion of humours* that nature produces the living phenomena; but that is not to say that the *circulation of humours* is *life*, because this is but a simple instrument, . . . Similarly, excretions and secretions do not constitute life. They are really only its supreme and most immediate *instrument*, by means of which nature can reject whatever is improper or foreign to it, and retain and assimilate whatever is useful to it for the conservation of the body.[33]

Motion is also at the root of nervous activity. The phenomena of sensation occur through a tonic or very sensitive tensile movement in the delicate nerves of sensory organs.[34] In the composite body, the soul needs the body as much as the body needs the soul. "The soul cannot, indeed, have any sensation of a thing, and consequently any thought or knowledge with regard to a sensible present object without the intermediary of the sensory organs; it cannot furthermore effect an act or execute its will without the aid of corporeal organs."[35]

[31] Georg Ernst Stahl, *Paraenesis ad aliena a medica doctrina arcendum*, Halle, 1738, §XVII, p. 50; *Paroenesis* in *Oeuvres médico-philosophiques*, op. cit. note 15 above, vol. 2, p. 224.

[32] Georg Ernst Stahl, *Theoria medica vera*, Halle, 1738, pp. 200–201; *Vrai théorie médicale* in *Oeuvres médico-philosophiques*, op. cit. note 15 above, vol. 3, p. 43. We also read that the heterogeneous body is "condemned to corruption if abandoned to itself", in *Mixti et vivi*, op. cit., note 30 above, §§V–XV, pp. 68–75.

[33] Stahl, *Paraenesis*, op. cit., note 31 above, p. 46; *Paroenesis*, op. cit. note 31 above, p. 169.

[34] Stahl, *Theoria medica vera*, op. cit., note 32 above, pp. 421, 438; *Vrai théorie médicale*, op. cit., note 32 above, vol. 3, pp. 428, 456–457.

[35] Ibid., p. 210 in the Latin text and pp. 46–48 in the French one. The same idea is expressed in *Disquisitio de mechanismi & organismi diversitate*, Halle, 1738; *Mécanisme et organisme*, in *Oeuvres médico-philosophiques*, op. cit., note 15 above, vol. 2, pp. 219–226.

One of the most difficult notions to comprehend in Stahl's work is his assertion that the soul acts wisely and consciously in the body. How then does one account for the fact that most of the body's functions are unconscious and automatic? Stahl's reply was that all activity is conscious in a newborn infant. It ceases to be so largely when it becomes habitual.[36] It was surely the least persuasive part of his hypothesis. At least mechanists had avoided that particular difficulty when they separated the psychic functions from the automatic ones.

Whatever the limitations of his conceptual system might be, Stahl paved the way for a revision of the fundamental principles of physiological theory. Although it was not obvious early in the eighteenth century, contemporary mechanical descriptions of living processes were too limited to provide a basis for continuing progress in the life sciences. Stahl's approach, on the other hand, paved the way for the production of an autonomous science of man.[37] Anti-mechanism was the proverbial idea whose time had come. Perhaps it was even somewhat ahead of its time, for when Stahl's work first appeared, it was largely dismissed or disregarded. Boerhaave simply denied that animism had anything to do with medicine, while Hoffmann preferred his animal spirits, subtle fluids, indwelling organizing forces, or, apparently, whatever else he needed to avoid the problems posed by a ubiquitous soul.

Animism first took root in the south of France among the medical faculty of the University of Montpellier who were, like their counterparts elsewhere, mechanistically inclined. Perhaps because of their school's own considerable reputation, students from Montpellier did not flock to Leiden in the same numbers as did those from elsewhere. Boerhaave's biographer, Lindeboom, suggests that the great man's reputation was less in France than elsewhere,[38] permitting his ideas to be more easily displaced there. In any case, once established, animism set Montpellier's physicians off in directions different from those dominating other schools. Even when animist notions had been displaced by organicist theories, Montpellier physicians continued to recall Stahl as an intellectual prophet, crediting him with the immeasurably important task of freeing physiology from iatromechanist excess.

Animism, or at least a mechanism greatly modified by animist notions, was introduced into Montpellier by *François Boissier de Sauvages* (1706–67), a flamboyant and effusive man upon whom even the heavens smiled. He was born, we read in a eulogy, at the precise time of a total eclipse of the sun. Whether or not this was propitious, Sauvages in time acquired a considerable reputation as a physician and a teacher, both in France and abroad. Eight years after receiving a doctorate from the University of Montpellier, he obtained, in 1734, a position as a *médecin survivancier* to a member of the faculty. As such, he was expected to assist his professor and to fulfil his professional obligations whenever necessary. In the eighteenth century, getting a professional position at Montpellier was a mercenary business. As often as not, the man who occupied a well-paid chair spent most of his time practising medicine lucratively elsewhere, while his *survivancier* performed all his professional

[36] Stahl, *Mechanismi & organismi*, op. cit., note 35 above, pp. 118–119; *Mécanisme et organisme*, op. cit., note 35 above, pp. 338–339.

[37] Roger, op. cit., note 7 above, p. 430; Duchesneau, op. cit., note 29 above; and Duchesneau, op. cit., note 2 above, pp. 1–31.

[38] Lindeboom, op. cit., note 25 above, pp. 355–374.

duties. In 1740, on the recommendation of François Chicoyneau, first physician to the king, Sauvages was appointed to a lectureship in botany. But it was understood that he would remain in the position only until 1758, when Chicoyneau's grandson would reach the age of twenty-one, at which time he would take up the position. Indeed, even though Sauvages became Royal Professor of Botany in 1751, he gave up the position, as expected, in 1758. Nevertheless, he remained associated with Montpellier for his remaining years.[39]

As a student, Sauvages would have listened to the lectures of such iatromechanists as Jean Astruc, Antoine Didier, and Pierre Chirac. About 1737, however, he began to adopt an animist viewpoint in his lectures. Many student theses written under his direction after this date tend to display a pointedly animist viewpoint. His new orientation did not go unnoticed by his colleagues. Bordeu, who was a medical student at the time, reported that the heated arguments lasted some six or seven years.[40] In time, however, the new approach triumphed, and Montpellier became so proud of its theoretical innovation that it too became a stultifying orthodoxy. In the middle of the eighteenth century, however, it was liberating. Its assimilation by persons hitherto iatromechanical is perhaps a tribute to Sauvages' persuasiveness as well as to the limitations perceived in the older theory.

This is not to say that Sauvages turned his back on mechanical principles. In all his numerous rambling works on many subjects he consistently affirmed that a physician ought to understand and to borrow the principles of philosophy, mechanics, mathematics, physics, and geometry, for they permit one to determine the capacity of vessels, the strength of solids, and so on. "I am persuaded", he wrote, "that one can establish no certain theory of the animal economy without the knowledge of Physics and Mathematics and that a man who will unite to the knowledge of these sciences those of Anatomy and *materia medica* will be more able to practise medicine with success than another man."[41] While he did not wish to return to pre-mechanistic conceptions of the living world, however, he denied that mechanism and physics were in themselves sufficient to study it.

Mechanistic explanation dominated a *Dissertation sur la nature et la cause de la rage*, commissioned by the French government. The blood of a rabid animal, wrote Sauvages, is observed to become sticky, because rabid venom contains "volatile alkaline particles" which coagulate blood. In the first stage of the disease, he wrote, heart contractions will become weaker with such blood. Its decreased quantity and velocity reduces its friction in the vessels, thereby forcing down the animal's temperature and producing a muscular lassitude. This weakness is countered by an increased activity on the part of the nervous system. Nerves are insinuated with the rabid venom, increasing the elasticity of their fluid and hence the victim's general

[39] Two largely biographical studies are: Étienne Hyacinthe de Ratte, 'Éloge de Monsieur de Sauvages', in *Les chefs d'oeuvres de Monsieur de Sauvages*, 2 vols., Lyons, V. Réguilliat, 1770, vol. 1, pp. xxv–lxxxvi; and Louis Dulieu, 'François Boissier de Sauvages (1706–67)', *Rev. Hist. Sci. applic.*, 1969, **22**: 303–322.

[40] Théophile de Bordeu, 'Recherches anatomiques sur la position des glandes et leur action', in *Oeuvres complètes*, 2 vols., Paris, Caille et Ravier, 1818, p. 204. Also mentioned by Alexis Alquié, *Précis de la doctrine médicale de l'École de Médecine*, 4th ed., Montpellier, Frères Ricard, 1846, pp. 19–23.

[41] François Boissier de Sauvages, 'Discours préliminaire', *Nosologie méthodique ou distribution des maladies en classes, en genres, et en éspèces*, 10 vols., Lyons, J. M. Bruyset, 1772, vol. 1, pp. 73–83

sensibility. A person in this condition tries to shut out all impressions, which by now seem intolerably strong. He is hydrophobic because his throat is very sensitive even to the water which he craves desperately. As the disease intensifies, violent muscular force produces insomnia, copious perspiration, an accelerated pulse, and general over-activity. And so he continued, always mechanistically.[42]

Iatromechanism alone, however, was insufficient to account for the action of drugs. A *Dissertation sur les médicamentes* of 1752 provides an illustration of Sauvages' particular modification of mechanism. All ingested products including drugs, food, and poisons are first broken down mechanically into their parts or molecules. One large pill, he wrote, acts ten times more slowly than a thousand small ones because the latter have ten times the surface area of the former. A drug that acts selectively upon the stomach and intestines does so simply because its particles do not pass into lacteals or other small vessels. Mercury acts upon the head because its molecules, heavier than blood, enter the straight carotid artery rather than turning the angle of the aorta and flowing towards the limbs, and so on. With a multitude of examples to back him up, Sauvages claimed that drugs act "by means of mechanical principles such as mass, velocity, the structures of the parts, calibre size and so on".

The significant point for our purposes, however, is that "Drugs act, not on a pure machine, but upon an animated one". A tiny bit of "tabac d'Espagne", Sauvages wrote, dissolves when it touches the membranes of the nose, producing violent sneezing. A grain of emetic causes the stomach to lift a column of water nearly two feet with a force calculated to be equivalent to thirty pounds of water falling from the same height. Since machines do not multiply forces, these minuscule causes must produce their powerful effects because they act through a living motor.[43]

He discussed the nature of the living motor at length in the preliminary discourse to his *Nosologica methodica* of 1769, in which he classified illnesses in imitation of Linnaean categories. "Man", he wrote, "is an aggregate or a being made up of a *living* soul and a *mobile* body." The former is the critical motor (*puissance motrice*) which moves the normally inert bodily mechanism. He claimed that in a cadaver, such as a recently drowned man, the mechanism is intact. We find therein all the faculties common to plants, which he believed act purely out of mechanical necessity, and to hydraulic machines. The fluids are the same. It is dead because there is no soul.[44] "Daily experience shows us that the soul is the principle of understanding, of appetite, of muscular motion, of the motion of the heart and of respiration, because these motions persist so long as it is present and cease when it is absent."[45]

A single substance, according to Sauvages, the soul possesses three faculties – it knows, it desires, and it moves. Their distribution and complexity vary between man and the animals. For example, the first faculty which is the source of sensation, imagination, and memory contains, only in man, the capacity for reflection, reason, and abstraction. Whereas all animals experience desire, it is modified by the will in

[42] Sauvages, 'Dissertation sur la nature et la cause de la rage', *Chefs d'oeuvres*, op. cit., note 39 above, vol. 1, pp. 15–105.

[43] Sauvages, 'Dissertation sur les médicaments', ibid., vol. 2, pp. 1–97. The quotation is from pp. 9–11.

[44] Ibid., pp. 9–11. Also *Nosologie méthodique*, op. cit., note 41 above, vol. 1, pp. 168–216.

[45] Ibid., p. 219.

humans alone. So it is, wrote Sauvages, that the surgeon depends upon his patient to volunteer to the knife and the criminal walks to the gibbet. The third or motive faculty is the source of all bodily functions whether these are voluntary or involuntary.[46] By resorting to this subdivision, Sauvages was able to account for the variety of physiological response without multiplying motive principles.

Although the soul acts principally in the brain, according to Sauvages, it cannot be confined to it. This is confirmed by an excised heart beating, by the moving parts of a dissected lizard, and so on. Under the soul's control in the intact animal, such motions are attributed to remnants of the nervous fluid which mediates between the soul or the brain and the parts.[47] Like many others, Sauvages tried to solve the problems created by the theory of a nervous fluid by assigning it both material and immaterial properties. New work on the subject of electricity offered Sauvages new possibilities for speculating about its special nature. Like light, he wrote, a nervous impulse travels at least thirty times more quickly than sound. Nervous fluid may be a kind of "elemental fire", electrical in nature. Like electrical fluid, it is very finely material, permitting it to travel over fibres to the parts much more quickly than any other material known.[48]

Clearly, Sauvages cannot be strictly labeled a Stahlian. His commitment to animism was, I suspect, far less fundamental than Stahl's had been. Stahl became an animist because he was preoccupied with integrated and complex activities directed towards some goal, such as occur only in living bodies. He considered inorganic nature to be a world apart. Sauvages was somewhat more cavalier than Stahl in employing the notion of the *anima.* He called upon it as a merely convenient explanatory device, a kind of physiological unknown to be invoked whenever the possibilities of the mechanist approach were exhausted. It was the animal motor, the device that moves muscles and confers a mechanical advantage whenever one is needed. Indeed, the *anima* that Sauvages postulated is in many ways a somewhat uncomfortable explanatory device. Its role was far more nebulous and vague than the purely voluntary and rational one assigned to it by the mechanists. Neither was it as noble or powerful as that omniscient instrument which Stahl believed integrates, calculates, and consciously oversees every living act. It seems, in fact, that it existed largely to fill in the gaps which mechanism left in Sauvages' understanding.

A precocious young man, *Théophile de Bordeu* (1722–76), began to attend Sauvages' lectures shortly after he began teaching at Montpellier. Caught in the middle of the controversy aroused by his teacher's unorthodoxy, Bordeu assented to his criticism of the purely mechanical approach to life. He was not satisfied, however, with the animist alternative. He opted instead for a monist anti-mechanism which owed a great deal to the ideas of Helmont and of Francis Glisson, who had written at length on the subject of the irritability and reactivity of living fibres.[49] Bordeu's approach was thoughtful, creative, and sufficiently novel to provoke great interest among his fellow physicians. More than any other physician, he was responsible for

[46] Ibid., pp. 216–255.

[47] Sauvages expressed interest in the animist notions of Robert Whytt, including his contention that the soul is extended throughout the body. For a discussion of this interest see French, op. cit., note 16 above.

[48] Sauvages, op. cit., note 42 above, pp. 54–71.

[49] See Chapter III.

discrediting iatromechanism and the Boerhaavian approach to medicine. Bordeu directed physiological investigation into a new and more productive direction which was to come into vogue in the middle of the century. He was to be undoubtedly the single most important influence on Xavier Bichat.

Bordeu was born in the south of France into a family whose men had been physicians for some four generations. He earned the title of *médecin-chirurgien* from Montpellier in 1744. Intensely ambitious and covetous of a specialization in the diseases of the rich, he left the south of France for the capital. To practise medicine in Paris, however, the diploma from Montpellier was not sufficient unless one enjoyed the patronage of a member of the royal family. Having no princely client, Bordeu wrote three theses for the Faculté de Paris, took an examination, and paid them a fee of fifty louis. By the time he reached the age of thirty, he was established at the Hôpital de la Charité. He is widely reported to have been a handsome, elegant, fluent, caustic, and somewhat metaphysically inclined young man whose charm and intellect brought him into the elegant circles of Paris. In the group that collected around Jeanne de Lespinasse, Bordeu met such *philosophes* as Denis Diderot and Jean d'Alembert, the editors of the *Encyclopédie.* Partly because of that association, his medical philosophy grew in reputation and influence in Paris and outside of France. Some of his most fundamental ideas found their way into the *Encyclopédie* and helped to shape the physiological viewpoint adopted in that widely read work.[50]

In his mature physiological system, vital function was associated with an active force of sensibility, which Bordeu believed resides in every animal organ. He considered that this sensibility responds to stimuli, thereby activating a force of irritability or contractility that always accompanies it. Together, they provoke the motions that constitute an organ's vital action. The notion that sensibility and irritability are fundamental physiological properties of the structural elements of the body had already coloured and shaped much of the work of Glisson and of others, including Helmont and Harvey. In the eighteenth century in general, these forces were to be subjected to very extensive study by many persons. I shall examine some of them in the next chapter. Here, however, I wish to consider the way in which Bordeu incorporated the phenomena of sensibility and irritability into his particular vitalist and organicist system.

Bordeu's 1743 doctoral thesis, the *Chilificationis historia*, offered its readers a non-mechanist theory of the secretion of the bodily juices. In it, Bordeu located the soul, which directs glandular and other activity in the nerves. The youthful thesis was expanded into the book which medical historians have widely judged to be Bordeu's most important contribution to physiology. *Recherches anatomiques sur la position des glandes et leur action* was published in 1752, soon after Bordeu became

[50] Anthelme Richerand, 'Notice sur la vie et les ouvrages de Bordeu', *Oeuvres complètes*, op. cit., note 40 above, vol. 1, pp. i–xxiv; E. Forgue, 'Théophile de Bordeu (1722–76)', *Les biographies médicales*, Paris, Baillière, 1937, pp. 97–128. The work of Sauvages and Bordeu, along with that of Whytt and Gaub, are briefly discussed with respect to their transition to vitalist models in Hall, op. cit., note 4 above, vol. 2, pp. 66–86. Théophile de Bordeu, *Correspondance*, ed. by Martha Fletcher, 4 vols., Montpellier, Université Paul Valéry, 1980, is insufficiently edited. Nevertheless, it reveals a great deal about Bordeu's personal life and the practical circumstances surrounding becoming a successful physician in eighteenth-century Paris, especially if one were from the provinces. The letters tell us nothing about Bordeu the scientist.

established in Paris. He lucidly expounded there his basic theory about the nature of life and the vital processes. The glands served as models for describing the functions of all the organs of a body. He carefully and decisively demonstrated that the traditional mechanist theories which treated glands as complicated sieves were simplistic and completely inadequate. Having done so, he went on to account for glandular secretion and excretion of humours, as well as of all other vital activity by referring to the sensibility that is part of their organization.

By studying glands, Bordeu was taking up one of physiology's most challenging questions. The body produces many different humours, all of which are critical to its wellbeing. Understandably, it was generally felt that if one could begin to comprehend the means whereby one gland is able to produce a secretion quite different from that of another gland, one could begin to fathom something fundamental about life's mysterious processes. Descartes and Borelli, in the seventeenth century, offered the simplest explanation of all. They visualized the glands as sieves in which situation, size, and shape of the pores determined what passed through them. But that hypothesis could really do little more than explain why large particles cannot pass into glands that have small openings. A considerable amount of important work on the anatomy and physiology of glands was subsequently done by other mechanists including Malpighi, Ruysch, Morgagni, and Boerhaave.[51] According to Bordeu, Boerhaave's theory was the one which was generally accepted in the mid-eighteenth century. Although he admitted that glands are more than mere sieves, he predictably still tried to understand their complex operations mechanistically:

> Hence the Humours will be various according to the Discharge of the Artery from the Heart, its Situation with respect to the Heart and the Trunk from whence it arises, its different Complications & Number of Ramifications before it terminates, the different Velocity of the Blood moving through it, the Proportion that the single Branch bears to its Trunk, the different propelling Forces, both external and internal, which discharge the Humours, the different Time of its standing in the common Cavity or Receptacle, & the various Passages it goes through, from whence making new Changes, with the different Degree of Absorption or Exhalation of the more fluid Parts from the secreted Juice; all these Causes, I say, concur to produce the vast variety of Humours observed in the several Parts of the Body, from that one common Mass the Blood, whose Particles are thus variously sorted, separated, & combined in as wonderful Manner.[52]

Once the glandular humour was thus secreted or manufactured, it had to be released by the gland into the body to carry out its functions. Boerhaave discussed the excretion process of the salivary glands in a work dealing with the larger question of digestion. He thought that the salivary juices are pressed out of their glands into the mouth by jaw movements during chewing. He presumed that this activity constricts certain muscles around a gland thereby compressing it. As the collected humour is expelled by the pressure, the gland is stimulated again to secrete more saliva from the blood.[53] It followed that lachrymal glands must secrete tears in response to a pressure

[51] Luigi Belloni, 'Boerhaave et la doctrine de la glande', in G. A. Lindeboom (editor), *Boerhaave and his time*, Leiden, Brill, 1970, pp. 69–82. Belloni analysed the mechanistic glandular theories of Malpighi, Ruysch, Boerhaave, and a few other mechanists. Boerhaave's particular conception of glandular function is also discussed in Lindeboom, op. cit., note 25 above, pp. 264–268; and in King, op. cit., note 25 above, pp. 59–93.

[52] Boerhaave, 'Of the different structure of the glands', *Academical lectures*, op. cit., note 26 above, vol. 2, p. 234.

[53] Boerhaave, 'Of the origin, nature and mixture of the saliva with the aliments', ibid., vol. 1, pp. 135–156.

around the eye, and so on. Albrecht von Haller, Boerhaave's most famous student, offered a similar explanation of glandular phenomena in his *First lines of physiology* of 1747.[54]

Though this tortuous striving after mechanical explanation did not commend itself to Bordeu, he admitted that Malpighi, Ruysch, Morgagni, and Boerhaave ought to be consulted on the subject of glandular activity. It was simply that they had not succeeded in accounting for its complexity. Bordeu set out to show first of all that the mechanist's approach was erroneous and then to substitute a more satisfactory vitalist explanation. Bordeu dealt first with the question of excretion of juice from the gland in which it had already been manufactured and stored. With painstaking skill he demonstrated that the anatomical space in which a gland is located can be neither altered in shape nor diminished in size. Nature has located glands in such a way that no muscles or other structures can press upon them. He referred to an experiment done eight years earlier in Montpellier, perhaps while he was preparing his doctoral thesis, in which a water-soaked sponge was inserted into a parotid gland cavity. No amount of jaw movement could extract even a drop of that water from it. Bordeu took his readers on an anatomical tour of the glands. Moving from the head downwards, he examined the lachrymal glands, the pituitary, thymus, thyroid, pancreas, suprarenals, and so on. He even examined the brain, though he pointed out that he did not himself believe it to be a gland. Malpighi and Boerhaave, among others, had so classified it because of its alleged production of animal spirits. In any case, in every one of the organs examined, Bordeu's evidence was quite decisive. Under normal circumstances, no gland is ever subjected to any physical pressure, so that the explanation for glandular activity must be sought elsewhere.[55] The evidence seemed to strike a decisive blow against mechanism.

How, then, can one account for glandular phenomena? Bordeu used the testicles and mammary glands as models to describe what occurs in the process of secretion in general. He believed that the sexual organs demonstrate clearly the combination of excitation, irritation, sensibility, and nervous action which are critical for an organ's function. The process of excretion proceeds in four steps. There is a spasmodic action of the gland or an erection; this produces a shaking or some other sort of irritation; the disposition of the excretory organ alters so as to facilitate humoral flow; and finally, there are some necessary changes in the vessels. He applied the model to the salivary glands, claiming that first, there is a kind of tickling around the parotid when one begins to eat, causing the gland to undergo a kind of erection or awakening; jaw movements provide friction, which is an irritation; the glands are prepared to excrete saliva, and finally, the vessels going to the gland allow more blood to collect around it and less to flow out so that more saliva can be manufactured. Lachrymal glands are irritated by air, inflammation, laughter, and coughing. He was convinced that in every gland, some appropriate irritation is always there to activate the force of sensibility, which is then responsible for excretion and for further production of the glandular

[54] Lester S. King, 'Introduction', to Albrecht von Haller, *First lines of physiology*, trans. and ed. by William Cullen, Göttingen, Wrisber, 1786, reprinted New York and London, Johnson Reprint Corp., 1966, pp. 109–134.

[55] Bordeu, op. cit., note 40 above, pp. 50–130.

humour. In summary, he wrote as follows: "Glandular excretion . . . does not proceed, as has been asserted, by the compression of glandular bodies, but by the *proper action of the organ*, action augmented by certain circumstances, such as *irritations, jolts*, and the *disposition of the vessels* of that same organ."[56] It is worth noting that though Bordeu maintained vigorously that physical glandular compression cannot be part of the excretory mechanism, he permitted a shaking and jolting to serve as the signal for irritability. If that is somewhat of a contradiction, it seems to have escaped Bordeu.

In the second stage of his investigation, Bordeu took up the question of how an organ secretes, that is to say manufactures, its humours. He considered that the force of sensibility was as important in this first stage of glandular activity as in the final one. Following on Boerhaave's observation that glands have more nerves in proportion to their bulk than do any other parts of the body, Bordeu remarked that the Creator surely would not provide so many nerves without giving them a function.[57] They are, he believed, the instruments that transport the critical signal of sensibility. The most compelling of six "proofs" advanced on behalf of their glandular role was simply that cutting the nerve supply to a gland stops its function. Another was that in sleep, a state assumed to be the result of relative nervous inactivity, the secretions of such as the salivary glands seem to be slowed. Bordeu neglected to mention, however, that the secretion of the digestive organs and kidneys are not so affected. He also thought it probable that nerves prepare a gland to make a kind of choice of appropriate material from the circulation. They are, therefore, the principal cause of the separation of the proper parts from the blood. In each orifice of each gland, there must be a particular type of small sphincter and several nervous fibrils which cause it to dilate or constrict depending upon the irritation of the nerves. The choice thus made by the organ is governed by what Bordeu called "a type [espèce] of *sensation*".[58]

Bordeu assumed that a similar indwelling sensitivity exists in every organ of the body. Nervous fibres are its instrument. It is unique in each organ, and therefore capable of performing its particular function. It is accompanied by an irritability which it activates, causing it to move the parts of the organ.[59] Sensibility is, therefore, the force that directs all organic functions. It even dominates in illnesses and directs the action of remedies.[60]

Because microscopy had revealed nervous filaments to be solid structures, Bordeu rejected the theory of animal spirits to which so many physicians had assigned deplorably vague and even contradictory properties. He believed that nerves act by means of motion. All organs possess a basic vital tension that is imparted to them by the brain by means of nerves' motion. This motion derives first from the force of the blood due to heartbeat; hence, as long as the heart beats, the brain experiences the blood's motion. This general vital tension maintains muscle tone and, indeed, life in general. The sensation, irritability, and activity of the various organs derive from a

[56] Ibid., pp. 121–146.

[57] Boerhaave, op. cit., note 52 above, pp. 210–211.

[58] Ibid., pp. 196–201.

[59] Bordeu, 'Recherches sur le pouls par rapport aux crises', in *Oeuvres complètes*, op. cit. note 40 above, vol. 1, pp. 420–421.

[60] For example, see ibid., pp. 420–421; 'Recherches sur l'histoire de la médecine', in *Oeuvres complètes*, op. cit., note 40 above, vol. 2, p. 669.

motion in excess of that basic tone.[61]

Like the mechanists, Bordeu treated the soul as a rational and spiritual being, involved in every conscious function. "Other functions", however, "seem to depend upon purely vital sensibility without it appearing that the soul has any role at all to play."[62] However, the soul frequently modifies even the latter functions. For example, Bordeu specifically linked the soul's activity to the phenomena which the body experiences in emotions, and claimed that each emotion or passion is connected with an organ so that some emotions suspend digestion, others produce tears, and so on. These effects are mediated through nerves, although, like all conscious sensations, emotions originate in the brain.

Bordeu continued to develop his sensibility thesis over the years. In time, he came to see the living body as a collection of organs, each of which possesses its own particular and separate life. "The living body", he wrote in 1775, "is a collection of several organs which live in their own way, which experience more or less sensation and which move, act and rest in determined times . . . the general life (of the organism) is the sum of all the particular lives."[63] He went so far as to locate the centre for these disparate sensibilities in the region of the stomach. "We place there the seat, the end point, the support of almost all the body action, of almost all the sensations; the play and tumult of the emotions."[64] For the body functions as a whole, Bordeu discerned a total of three co-ordinating centres. "The brain, the heart and the stomach are . . . the triumvirate, the triple support of life . . . the three principal centres from which consciousness and motion flow and to which they return after having circulated."[65]

The similarity between these latter notions and Helmont's visions is more than mere accident. Bordeu wrote of Helmont in superlatives, crediting him perhaps too enthusiastically with the overthrow of Galenism and the creation of a new medicine. This lends weight to the contention that Bordeu's concept of general sensibility was fashioned after the grand *archeus* and the proper sensibilities of the specific organs were based upon the individual *archei.*[66]

It is interesting that in his search for alternatives to the prevailing physiological systems, Bordeu should look to the vision of a man who had challenged orthodoxy over a century earlier. Bordeu's language, of course, sits more comfortably with us than that of Helmont. Whereas we can cope with "sensibility", the vague and mysterious *archei* are thoroughly fanciful. Nevertheless, it was there that Bordeu found the idea that one must search for the life of a body in its own parts.

Bordeu's physiological system had considerable influence upon his own century. He was seen as the most eminent representative of what became known simply as the

[61] Bordeu, op. cit., note 40 above, pp. 196–201.

[62] Bordeu, op. cit., note 60 above, pp. 667–669.

[63] Bordeu, 'Recherches sur les maladies chroniques', in *Oeuvres complètes*, op. cit., note 40 above, vol. 2, p. 829. For an interesting examination of the psychological consequences of this physiological theory, see Paul Hoffmann, 'L'idée de la liberté dans la philosophie médicale de Théophile de Bordeu', *Studies on Voltaire and the Eighteenth Century*, 1972, **89**: 769–787.

[64] Bordeu, op. cit., note 63 above, p. 801.

[65] Ibid., pp. 829–831.

[66] The connexion between the notions of Helmont and of Bordeu was made by Marie-Jean-Pierre Flourens, 'Coup d'oeil historique sur l'étude analytique de la vie', *De la vie et de l'intelligence*, Paris, Garnier Frères, 1858, p. 51.

Montpellier school. One of its representatives, Dr Ménuret de Chambaud (1733–1815), wrote more than forty articles for the last ten volumes of the *Encyclopédie*, thereby using this widely read work to propagate the new physiology. He conceived of a motor faculty regulating unconscious vital motions. It was equated with Glisson's irritability, which was taken to be synonymous with "sensibility", "mobility", and "contractility". Diderot commissioned another Montpellier professor, Henri Fouquet (1727–1806), to write the articles on 'Sensibilité' and 'Secretion' for the *Encyclopédie*. Fouquet tended to regard sensibility as a kind of sensitive soul, rather more a substance than a property of matter. Most of his other ideas, however, including those having to do with secretion, faithfully followed Bordeu.[67] Many other people, including Bichat, examined many of these notions over the years as they became part of the language of medicine.

In spite of Bordeu's considerable impact on the medical world in general, his views were eclipsed in Montpellier by those of *Paul-Joseph Barthez* (1734–1806), who was to give the vitalist theory still another form. Barthez did not share Bordeu's admiration for Helmont, and he dismissed the numerous specific organic sensibilities as a vain multiplication of causes. He preferred to attribute the functions of life to a single principle. That principle, however, was different from the soul or *anima*.

Barthez was born in Montpellier and received a medical doctorate from that university in 1753. After serving some years as a physician and a medical consultant to the French army, he applied for the chair of medicine at Montpellier in 1759. When he finally received the position two years later, he joined a faculty of men whose professional relationships were most remarkably and openly petulant, selfish, and even vindictive. Few things aroused their outrage and indignation so much as having one of their number displaced in an academic procession by a man of junior rank. Barthez created considerable animosity because he served as *survivancier* to Jean-François Imbert, who was chancellor of the university from 1761 to 1785. Whenever Imbert was absent from the university, Barthez enjoyed all the privileges of his rank. When Imbert returned, Barthez' status reverted to that of a junior instructor. Barthez and his colleagues alternately played the situation for what it was worth so that he became a kind of social yoyo. Remarkably, Barthez thrived and prospered in that atmosphere. Various people who knew him have described him as a short and ugly man with an angry, vindictive, ambitious, and jealous nature,[68] but he was also an excellent teacher with a prodigious memory and a remarkable capacity for work. It was a combination of qualities which earned him a reputation unsurpassed at the University of Montpellier to this day.

Over the years, he became physician to the Duc d'Orléans, chancellor of the University of Montpellier, consulting physician to the king, a member of the National Health Council, and finally, a State Councillor. He spent most of his time in Paris, leaving his duties, meanwhile, to *survivanciers*. He passed the revolutionary years in

[67] The Montpellier physicians are discussed in relation to Diderot by Roger, op. cit., note 7 above, pp. 619–641. Ménuret, Fouquet, and Lacaze in particular are discussed on pp. 631–641.

[68] A young student described Barthez in 1758 as a man "mal bati, de vilaine figure, sourd d'un côté, entendant mal de l'autre, ayant de mauvais cheveux mal plantés, done cependant il est amoureux, en habit noisette à boutones d'argent". Quoted by Alice Joly in 'Un étudiant lyonnais à Montpellier au XVIIIe siècle', *Monspel. Hippocr.*, 1970, **13**: 11–23.

Narbonne, presumably avoiding the capital because of his favoured status in the former regime. In 1801, he was named a member of the Institut National de France, and thereby restored to the profession. He served as a consultant physician to Napoleon under the Empire and in 1802, he returned to the faculty of Montpellier.[69]

Barthez' memory is today much revered at Montpellier. His larger-than-life statue, erected in 1856, stands guard at the entrance to the school. A street and a reading room are named after him, and his bust looks down in various parts of the school. More importantly, his contribution to medical philosophy is still frequently alluded to by members of the medical faculty. In 1971, for example, one wrote that "The essential ideas which he uttered have today become so evident that they are contested by no one. Only the history of medicine is there to recall that it was not always so."[70] If asked to account for his enthusiasm, the writer of those words would be unlikely to claim that life is linked to a vital principle which may be material in nature and whose forces transcend physical ones. But just such a principle was Barthez' preoccupation in his work, especially in the *Nouveaux éléments de la science de l'homme.*

In Paris, Barthez was a close friend of Jean d'Alembert, an exponent of Newtonian philosophy. Barthez is alleged to have acquired the germ of his mature ideas about life from him.[71] Indeed, Barthez also claimed for himself the medical mantle of Newton, believing that he was working in accordance with the great man's principles. The object of natural philosophy, he wrote in the preliminary discourse, is research into the causes of natural phenomena. Immense progress has been achieved in astronomy, for example, because a multitude of observed effects came to be assigned to a few experimental causes such as gravity. That is the essence of scientific progress, which consists in the establishment of more and more general laws that account for more and more phenomena. Thus the number of causes necessary to explain phenomena diminish progressively. To emulate Newton was, for Barthez, to search for forces or causes that would account for a multitude of physiological responses as coherently as the gravitational force explained physical and astronomical ones.

The "good philosophical method" had eluded the ancients, Barthez wrote, because they were too apt to multiply causes. More recently, other persons had erred in the opposite direction. For example, the mechanists tried to relate all animal activity to the force of impulsion while the animists resorted to a single ubiquitous soul. Most interestingly, Barthez also warned his readers against the new sect of the "solidists". Neither mechanists nor animists, its members attached great importance to Helmont's notion that each organ has a life proper to it, in addition to the separate particular life belonging to the entire body. Unmistakably, he was thinking of Bordeu and of other similarly-minded members of the Montpellier school. He took their work to be a form of misleading and pointless multiplication of motive forces.[72]

Barthez sought to replace these various flawed theories with his notion of the vital

[69] The biographical material is taken largely from Louis Dulieu, 'Paul-Joseph Barthez', *Revue Hist. Sci. applic.*, 1971, **24**: 149–176. A complete bibliography of Barthez' published works appears at the end of the article.

[70] Ibid., p. 170.

[71] Flourens so argued in op. cit., note 66 above, p. 83.

[72] As well as the members of the Montpellier school, Barthez must have had in mind the works of Francis Glisson, Julien Offray de la Mettrie, and Denis Diderot.

principle, which he regarded as the most general of experimental causes. He believed that it coexists with the body and the rational soul and that it possesses both sensitive and motor components. The respective functions and the inter-relationship of body, soul, and vital principle were the main subject-matter of the *Nouveaux élèments*.

Barthez used the word "principle" to denote simply an experimental cause of motion. Nature employs several of them, the most simple being *impulsion*, whereby an object is moved by physical contact. The slightly more complex force of *attraction* accounts for gravity, for particular affinities, and for the phenomena of magnetism and electricity. Still more complex forces form ice crystals and angular crystallized salts. The highest and most complicated force in nature is the *vital principle*, which is found in plants and animals and whose functions cannot be accounted for more simply.[73] It is more complicated in man than in animals, and least complicated in plants.[74]

Barthez did not, however, claim the discovery of the vital principle distinct from both body and soul for himself. He recognized it in Francis Bacon's claim that man has an irrational, corporeal soul in addition to his rational, spiritual one. Albeit "carried away by ideas approaching delirium", Helmont had glimpsed the truth when he claimed that man is endowed with a principle of life distinct from both body and soul. More recently, Hoffmann had speculated that the life principle occupies a place between that of body and soul. Even some physicians from Boerhaave's school, and Barthez especially had in mind Gaub, Boerhaave's successor at Leiden, attached a vital principle of some sort to the body.[75]

Barthez appears to have adopted the theory of a vital principle simply because he was uncomfortable with the idea that the source of motion of a body resides in its matter. On the other hand, he turned his back on animism because he could not imagine that the soul, an instrument of the will as he understood it, can participate in automatic activity. It has no sense of internal vital motions and it can neither suspend nor modify the motion of, for example, the heart and intestines. Nor can it account for the frequent conflicts an individual experiences between his will and his appetites. Recall here Sauvages' examples of the criminal walking unaided to the gibbet and the patient to the operating table. Sauvages had dealt with this conflict by subdividing the soul into faculties that presumably spar with one another. Finally, Barthez pointed out that animism cannot account for the ability of parts recently separated from a body to move and to respond to stimuli. He offered the examples of the decapitated chicken and the extirpated heart, and claimed even to have seen movement in the head and trunk of a decapitated man. He thus opted for a living principle distributed throughout the body.[76]

[73] Paul-Joseph Barthez, 'Discours préliminaire', in *Nouveaux élèments de la science de l'homme*, Montpellier, J. Martel ainé, 1778, pp. i–xxviii. For a discussion of Barthez' views concerning the nature of the vital principle, see Réjane Bernier, 'La notion de principe vital de Barthez', *Archs. phil.*, 1972, **35**: 423–441; and Elizabeth L. Haigh, 'The vital principle of Paul-Joseph Barthez: the clash between monism and dualism', *Med. Hist.*, 1977, **21**: 1–14. The work of Bordeu is discussed, together with that of Barthez, by Duchesneau, op. cit., note 2 above, pp. 361–430.

[74] Barthez, op. cit., note 73 above, pp. 1–6.

[75] L. J. Rather, *Mind and body in eighteenth-century medicine, a study based on Jerome Gaub's De regimene mentis*, London, Wellcome Institute, 1965.

[76] Barthez, op. cit., note 73 above, pp. 28–35.

What then is the nature of this vital principle? Does it have some centre or focus? If the extirpated heart was not beating as a consequence of either an indwelling sensitivity or a soul, the only other explanation was that it must retain some part of the vital principle within it. The principle, therefore, must be divisible. The force exists, Barthez wrote, in solid and fluid parts of the body alike. It is, for example, the source of the motion of the blood.[77] The vital principle is endowed with both sensitive and motor forces which interact to produce the complex phenomena that we observe as life. Barthez distinguished between sensibility and irritability. Although each organ possesses at least minimal sensibility, the force is most abundant in the brain and the nerves. Meanwhile, all the solid parts of the body possess some amount of the motor force. In structures such as bones, however, it exists only to nourish the parts and to effect regeneration if the bones should be damaged.[78]

In all cases, Barthez believed, there is a connexion between the two forces of the vital principle such that mobility depends upon sensibility. The extirpated heart, for example, continues to beat because some of the vital principle remains in it and "this part, when these members are irritated, is caused to move by the sentiment it has of an irritation". Sensibility, therefore, is not necessarily conscious. But by so speculating about the existence of unconscious sensibility, Barthez was moving perilously close to the position of Glisson and of Bordeu. Indeed, he wrote of the sensibility proper to each organ which exists apart from the nerves. Its existence is particularly clear in the case of certain small animals such as shellfish, which have a sensibility in spite of having no nerves at all. But that would be clearly our interpretation rather than Barthez' own.

The motions of the bodily parts, Barthez wrote, are of two sorts. The first is very rapid while the second is too slow even to follow. The former occurs in the motion of the muscles, the iris, the intestines, the heart, and so on. It is impossible to account mechanically for the enormous forces which the vital principle exercises in the contraction of various muscles. For example, the chewing of bones, the strength of someone suffering convulsions. the lifting of heaving weights, and so on, would break or tear those same bodily parts if the supra-mechanical life force were absent. The slow motions, on the other hand, are tonic forces that exist in all the parts, providing a degree of cohesion which external agents tend to weaken. One observes them in blood vessels, in the cellular tissue of the skin, in voluntary muscles, and wherever a basic tone must be maintained. Weakened during serious illness, they are also strengthened by exercise, which forces an increase in the cohesion of the muscular parts. So it is that the body can literally and properly be said to fall apart at death.[79]

The vital principle moves the humours of the body including the blood, and it maintains a suitable degree of heat in the body by agitating its solids and fluids.[80] Finally, it maintains the unity of the body by means of sympathetic connexions between the parts. The voice changes in puberty, for example, in response to changes in the sexual organs; abscess of the liver frequently follows a wound in the head;

[77] Ibid., pp. 101–117.
[78] Ibid., pp. 42–63.
[79] Ibid., pp. 64–100.
[80] Ibid., pp. 118–141.

inflammation in one eye will produce the same symptoms in the other; if the pupil of one eye responds to light, the other will follow suit, and so on. Barthez labelled these "sympathies", and alleged that they travel by means of nervous pathways. There is a perpetual reciprocal communication between the tonic forces of the nerves and a constant antagonism that keeps these forces in equilibrium.[81]

It might well be objected that so far, Barthez had claimed nothing which would make it necessary for him to reject the basic arguments of other members of the Montpellier school. After all, both he and Bordeu believed that irritability and sensibility are distributed in all the organs and parts. At one point, Barthez even wrote of the vital principle acting on what he called the molecules, that is to say, on the smallest parts of the muscular fibres so as to disperse them and to bring them together again.[82] Why not, then, simply say with Bordeu that a vital motion is a consequence of the sensibility resident in the moving part? As we have seen, Barthez apparently did not do so, insisting instead upon a separate vital principle, simply because of his conception of matter as inert substance.

He did, however, offer other bits of evidence in support of the assumption that the body and its motive force are separate entities. In so doing, he began to sound very much like Sauvages. He remarked, for example, that life and consequently the vital principle can be destroyed with no alteration of the integrity of the body, using Sauvages' example of the recently drowned man whose body is still intact. On the other hand, Barthez went on, the vital principle can survive a long time, even with substantial lesions of such essential organs as the heart and brain. Barthez also remarked on what he called a "pre-established harmony" which exists between the vital principle and motions in relation to organs that do not yet exist. For example, a small bird will make flapping motions when its wings are still too feeble to support it, and a calf will playfully butt with non-existent horns. Barthez took these examples to be evidence of the vital principle's independent existence, rather than simply being a modality of the living man.[83]

Barthez conjectured that the vital principle might be material in nature. In fact, many of his assumptions concerning its behaviour were only consistent with such a belief. He thought, for example, that it exists in a latent or concentrated state in pupae and birds' eggs, where it awaits animal heat to form and give life to an organism. He wrote in various places of lesions of the vital principle and of poisons acting to destroy it. As one commentator observed, one cannot concentrate, produce lesions in, or poison an immaterial principle.[84]

Barthez remarked that muscular motion is extinguished when nervous connexions to muscles are severed or when blood vessels are ligatured. Other organs such as intestines or the heart, however, are more independent and durable. It must be, he thought, because the latter, having more of the vital principle within them, depend less upon sympathetic connexions with other organs. Barthez tied this alleged quantity of the vital principle to the phenomenon ot sleep, reasoning that it is the organs with the

[81] Ibid., pp. 142–172.
[82] Ibid., pp. 68–69.
[83] Ibid., pp. 27–41.
[84] Bernier, op. cit., note 73 above.

greatest proportion of vitality that require least rest. Considering sleep to be the repose of muscles, he thought that it is produced by alterations in the brain. Such organs as the heart, arteries, respiratory muscles, and so on, must continue to act without respite, however. Hence the larger amount of the vital principle allotted to them.[85] A relatively trivial speculation by Barthez, this theme was subsequently developed by his student Jean-Charles Grimaud. Twenty years later, Bichat, in turn, considered the nature of sleep and in imitation of Grimaud, used it to develop his crucial division of life into animal and organic categories.

Finally, Barthez discussed death, defining it as the irrevocable cessation of sensibility and of vital motions. His uncertainty about the precise nature of the vital principle, however, prevented him from knowing its fate when life leaves the body:

> If this Principle is only a faculty united to the living body it is certain that it perishes with the body. If it is a being distinct from the body and the soul, it may perish outside the extinction of its forces in the bodies which it animates; but it may also pass into other human bodies and vivify them. . . . It is possible that the end of the vital principle is related to its origin. Thus in supposing that it may have emanated from a principle which God created to animate the worlds, it may be rejoined to the Universal Principle at death. . . . Whatever the destiny of the vital principle of man is at death, when his body returns to the earth, his soul returns to God who gave it to him and who assures him immortality.[86]

Believing his vital principle to be capable of unifying existing knowledge and of directing future observation and experiment, Barthez compared its alleged discovery to Newton's discovery of universal gravitation. He was convinced that his vital principle distinguished his system from all previous ones. Nevertheless, it is clear to us that it does not belong on the same ontological plane as, say, Newton's laws of motion and that, in fact, Newton would have consigned the vital principle disdainfully to the scientific rubbish heap of hypotheses.

The naturalist Georges Cuvier put his finger on the difficulty attached to postulating the existence of the vital principle. He pointed out that gravity is defined precisely so that the motion of bodies towards each other is described in accordance with a specific law. The vital principle, on the other hand, was described only in the most general terms. Furthermore, he questioned the validity of postulating a system which is neither material nor immaterial, neither mechanical nor intelligent. To say that the phenomena of muscular contraction, sensibility, curing of wounds, formation of the foetus, reproduction of the species, and so on, are all effects of a simple, single principle is merely to enumerate phenomena, not to explain them.[87]

Although we must concur with Cuvier's criticism, it is important not to lose sight of the intellectual authority that Barthez wielded, especially in Montpellier where his ideas commended themselves to generations of students. One of his students, who is important for our purposes because he assists us in understanding Bichat, was *Jean-Charles Grimaud* (1750–89). When Barthez became chancellor of the University of

[85] Barthez, op. cit., note 73 above. pp. 234–244.

[86] Ibid., pp. 347–348.

[87] Georges Cuvier, 'Histoire de la classe des sciences mathématiques et physiques', *Mém. Inst. Nat. Sci. Arts: Sci. math. phys.*, 1806, **7**: 1–79. A translation of pp. 76–79 is found in William Randall Albury, 'Experiment and explanation in the physiology of Bichat and Magendie', *Stud. Hist. Biol.*, 1977, **1**: 105–106. See also, Georges Cuvier, 'De Barthez, de Médicus, de Desèze, de Cabanis, de Darwin et de leurs ouvrages', *Histoire des sciences naturelles*, 5 vols., Paris, Fortin, 1843, vol. 4, pp. 27–46.

Montpellier in 1785, Grimaud was awarded his anatomy chair. That year, the Imperial Academy of Sciences in St Petersburg invited the medical community of the world to consider the following question:

> Since nutrition and growth occur in various parts of the animal body which have no vessels such as the epidermis, nails, hair, hooves as well as those with a small number of vessels like bones; as well as the rapid growth of the embryo at a time before there is a heart or vessels, it must be that there is a force proper to the animal substance which brings nourishing juices to all the parts in proportion to their growth. . . . What is this force? Is it the same as that of attraction or is it properly and uniquely due to the living substance of animals and plants? Are there perhaps different forces . . .?

It was a particularly suitable challenge for persons who had studied at Montpellier, and Grimaud duly sent a memoir to the Russian academy. A version was later published as a *Mémoire sur la nutrition* in Montpellier in 1787.

At the beginning of the work, Grimaud divided animal functions into internal and external, meanwhile warning his readers that the division is merely a device to assist us in ordering our ideas. By means of its external functions, an animal encounters the surrounding world. It "moves out of itself, it extends itself, it enlarges its existence, it carries and shares it with the objects which surround it; it studies these objects." This is achieved by means of the senses and with locomotion, all of whose activities are controlled by the brain.[88] Internal functions, including nutrition, glandular secretion, respiration, and reproduction, maintain the organism. In imitation of Bordeu, he speculated that they are controlled from a sensory centre at the superior orifice of the stomach.

Grimaud, then, proceeded to distribute the sensible and motor forces between the internal and external functions. An animal motor force (*force motrice animale*), he wrote, produces external functions by agitating fibres; it is paralleled in the internal life by the vital motor force (*force motrice vitale*) whose activity completely escapes consciousness and over which the will exercises no control. Its principal object is to distribute nourishing juices to all the parts, and it is subordinated to an internal vital sense (*sens vital intérieur*) whose activity is also apart from consciousness and the will.[89]

In a second *Mémoire sur la nutrition* and in a *Cours complet de physiologie*, Grimaud discussed the two lives in relation to sleep. "Sleep", he wrote, is a state or "effort of the internal parts just as wakefulness is an effort of the external parts." The two states, and hence the two lives, dominate each other alternately. So it is that the state of the foetus and that of plants is one of uninterrupted sleep.[90]

Bichat's vital theory was rooted in the same distinction between external and internal functions which Grimaud described. He did not, however, treat them as merely a convenient classificatory device. Rather, he believed them to be a fundamental division rooted in nature itself.[91] Bichat's five vital properties, in fact, were based upon the animal-organic division. The correspondence between Grimaud's *force motrice vitale, force motrice animale*, and *sens vital intérieur* and Bichat's *organic contractility*,

[88] Jean-Charles Grimaud, *Mémoire sur la nutrition*, Montpellier, Jean Martel, 1787, pp. 2–3. See also, Grimaud's *Cours complet de physiologie*, 2 vols., Paris, Méguignon-Marvis, 1818, vol. 1, pp. 37–53.

[89] Grimaud (1787), op. cit., note 88 above, pp. 19–40.

[90] Ibid., pp. 140–155; and *idem* (1818), op. cit., note 88 above, pp. 195–210.

[91] Bichat, op. cit., note 1 above, pp. 1–9.

animal contractility, and *organic sensibility* respectively is unmistakable,[92] and there can be little doubt that Bichat was familiar with Grimaud's speculations.

Grimaud's answer to the problem posed by the St Petersburg Academy was a *digestive force* which belongs to the internal life and to which he attributed all the complex activity surrounding the ingestion and assimilation of food. He described its functions much like he did those of the *force motrice vitale*. It is not altogether clear, therefore, why he chose to distinguish between them. Bichat would dismiss this as the vain multiplication of forces, arguing, in a manner reminiscent of Barthez, that contractility and sensibility alone are sufficient to account for all living phenomena.

Other people, however, were even more guilty than Grimaud of multiplying forces, faculties, and principles. By the latter part of the eighteenth century, traditional mechanism had been superseded by the imagery of active tissues whose activity is not reducible to that of the non-living world. By the 1890s, Montpellier was unequivocally a vitalist school, and it was becoming complacent about its achievements. Perhaps it was a sense of confidence which lulled some of its younger faculty into a lazy use of language. An example is *Charles-Louis Dumas* (1765–1813), who obtained a medical degree from Montpellier in 1784, and acquired considerable skill and a reputation to match. A brief glance at Dumas' discussion of physiology demonstrates how careless Barthez' successors became with his vitalist thesis so that it became a complicated and even unwieldy device that could not order one's thoughts. Although Dumas wrote on a wide range of subjects, his basic theoretical position is contained in a four-volume *Principes de physiologie* published between 1800 and 1803 when he was Dean of the Faculty of Medicine at Montpellier. After he had offered the obligatory credit to Helmont, Stahl, and Haller, Dumas praised Bordeu, Lacaze, Barthez, and Grimaud for providing a solid base upon which to deal with organic studies. Borrowing from these various persons at random, he only succeeded in confusing their notions utterly.

Dumas was not, for example, particularly concerned about the nature of the vital principle, considering it to be merely an "x" or unknown quantity somehow intermediate between the soul and matter. It has been variously labelled, he wrote, as a soul, *archeus*, vital principle, and so on.[93] Because of its presence, an organism lives longer than its parts, which are continuously undergoing alteration, absorption, and excretion. At times hidden in a seed or a cocoon, its dormant vitality is able to stir and to reassert its primacy over the physical and chemical forces of the larger world.[94]

In any case, Dumas was more concerned with vital forces than with their source. In clear imitation of his teacher Barthez, he wrote that elementary particles of organized matter possess all the inorganic forces such as gravitation, attraction, repulsion, and inertia.[95] To these, of course, are added the forces of sensibility and mobility. Dumas did not stop there, however. He also described an attracting force (*force assimilatrice*) and a force of vital resistance. Much like Grimaud's digestive force, the former oversees the transformation of food into bodily substance at the intimate level of tissues and organs. The latter was presumed to exist in order to retard corruption by prevent-

[92] Ibid., pp. 129–133.

[93] Charles-Louis Dumas, *Principes de physiologie*, 4 vols., Paris, Deterville, 1800–03, vol. 1, pp. 115–143.

[94] Ibid., pp. 159–248.

[95] Ibid., pp. 266–278.

ing certain fluids from coagulating, by keeping the stomach intact in spite of its solvent juices, and so on.[96] And so, not heeding his mentor's frequent warnings about multiplying forces, Dumas fell into the trap of enumerating phenomena without explaining them.

Dumas was, reasonably enough, uneasy with Grimaud's external-internal classification of functions on the grounds that all functions overlap. Instead of dispensing with it, however, he expanded it to correspond to the four faculties of his vital principle. With the first division, he wrote, the animal establishes a relationship with its surroundings; with the second it maintains organic solids and fluids in their particular state; with the third, nature conserves the body by means of nutrition, sanguification, and excretion; with the fourth, it invokes a reciprocal relationship of affections and thoughts between persons.[97]

Classifications, of course, need not be treated as absolute categories. They are more properly devices for ordering one's thoughts. Even allowing for that, however, Dumas' system is an unnecessarily complicated one, careless in its terminology and cavalier in its categories. This very unwieldy quality, however, testifies to the thoroughness with which his predecessors at Montpellier had done their propaganda work. Vitalism being by then axiomatic, it had become the framework within which to work.

In view of its considerable reputation before 1800, it is interesting that one encounters Montpellier virtually not at all in the nineteenth century. When the debate about the nature of the body and the means appropriate to study it were being vigorously pursued in Paris and elsewhere, Montpellier appears to have been unhealthily complacent. Its teachers, apparently, recycled what was by then a traditional form of vitalism, revering the memory of its great men and especially of Barthez. Student theses again and again invoked similar arguments on behalf of a unique vital principle. I admit that my impression may be false, based upon insufficient knowledge of the school after 1800. On the other hand, the statues, plaques, and eulogies produced in Montpellier in the last century suggest that persons there believed that their school had reached a peak in the work of Barthez.

Had it not been for the Revolution, Bichat would probably have followed in his father's footsteps and studied medicine at Montpellier. In any case, his ideas about the nature of the organism place him firmly within that school's tradition. Indeed, he was one of the last persons to insist, as a Montpellier physician might have done, that he was working within the context of the mechanist-vitalist debate. Writing on anatomy and physiology about the same time as Dumas, Bichat expressed far more coherent ideas. Nevertheless, the two men shared common intellectual roots. To Bordeu, Barthez, and Grimaud, Bichat owed many of his most compelling arguments on behalf of the existence of vital forces, their nature and their distribution. It is true that the ideas of the Montpellier school had become the stock-in-trade of physicians and philosophers at the time. Bichat's knowledge of their work, however, was so thorough as to suggest that their books were frequently at his elbow as he wrote and that what he was doing was basically elaborating and emending their notions. His achievement

[96] Ibid., vol. 1, pp. 313–348; vol. 2, pp. 132–153.
[97] Ibid., pp. 67–84.

can be summed up as a marriage between their explicit vitalism with notions of tissue reactivity and a Lockian epistemology, both of which subjects are discussed in subsequent chapters. He produced, thereby, a system to which virtually all persons conversant with physiology and anatomy assented in the first decades of the last century.

III

IRRITABILITY AND SENSIBILITY: THE FORCES OF LIFE

In his *Anatomie générale*, Bichat wrote "It is easy to see that vital properties are reduced to sensation and motion."[1] Most physiologists and physicians in 1801 agreed with him, and by referring all living activity to an ability to sense and to move, Bichat was adopting what had become a commonplace assumption. The two properties had been studied even in the ancient world. More importantly for our purposes, they had been investigated anew in the seventeenth century, and they had received particular attention in the preceding fifty years. Bichat's observations on animals had led him to conclude that there are, in total, five vital properties or subdivisions of sensibility and contractility.

In the mid-seventeenth century, Francis Glisson observed the reaction of fibres in response to a stimulus and made this capacity, known as irritability, the basis of his theory of bodily functions. We have already observed that sensibility preoccupied members of the Montpellier school a century later. Both Glisson and Bordeu believed that sensibility and irritability are intimately linked together in the body in such a way that the one always provides the signal which activates the other. Albrecht von Haller and Robert Whytt denied that they are tied together, claiming instead that they are distributed in various proportions in different organs and parts. That is, certain parts demonstrated considerably more sensibility to stimuli than others, which frequently behaved as though they possess more irritability or mobility. As we shall see, these differences between the various investigators had to do largely with the particular definitions they applied to the two properties. The notion of sensibility and irritability residing in the bodily parts provided Julien Offray de La Mettrie and Denis Diderot with the ingredients they needed to state a coherent materialist theory of living matter by which a soul was deemed unnecessary even to explain consciousness and rationality. The prominence of these various persons in the history of science and of medicine attests to the fact that a good deal of the progress made in physiological theory especially during the second half of the eighteenth century derived directly and indirectly from the study of sensation and motion.

To a large extent, this investigation stood apart from the mechanist-vitalist debate. The men who did the work in the eighteenth century nevertheless frequently declared their allegiance to one or other metaphysical position. Whytt, for example, was an animist though not, he insisted, a Stahlian; Haller had been a student of Boerhaave and, understandably, he considered himself to be a mechanist. The notion that sensibility and contractility exist in the bodily parts, however, was a particularly suitable one with which to defend an organicist or monist vitalism. For example, it suited Bordeu's and Bichat's polemical purposes admirably. Like the mechanists, the

[1] Xavier Bichat, *Anatomie générale appliquée à la physiologie et à la médecine*, 4 vols., Paris. Brosson, Gabon, 1801, vol. 1, pp. 99–101, 112–114. See also his *Recherches physiologiques sur la vie et la mort*, Paris, Brosson, Gabon, 1801, p. 130.

animists believed that the body behaves in accordance with mechanical principles but they differed over the origin or source of the body's motion. Bordeu and Bichat, on the other hand, simply denied that physical and mechanical laws had anything substantial to do with living activity. The separation of nature into organic and inorganic realms operating under two separate and distinct sets of scientific laws and principles were at the root of Bichat's vitalist theory. Like the Montpellier vitalists, he argued that organic sensibility and irritability were dominant over the universal physical properties of attraction and gravitation with which they coexisted.

Temkin has shown that such terms as "irritability", "irritable", "to irritate", and "to stimulate" occur in Galenic physiology. Galen himself wrote of a discharging mechanism in such parts as the gall bladder, stomach, intestines, urinary bladder, and uterus. Because they possess an irritable faculty, those organs are able to discharge their products without the intervention of the will or the consciousness.[2] William Harvey similarly used the notions of irritation and excitation in his study of the reproductive process.[3] The modern concept of irritability, however, is normally traced back to *Francis Glisson* (1598–1677), who generalized it in such a way that it came to be seen as an active force, an unconscious biological property spread throughout all parts of an organism.[4]

In his *Anatomia hepatis* of 1654, Glisson considered the irritability of the gall bladder; in the *Tractatus de ventriculo et intestinis* of 1678, he treated tissue reactivity as a generalized principle, an ability of an animal fibre to contract in response to a stimulus. In the earlier work, he linked it to the discharge of bile from the liver to the gall bladder and subsequently into the digestive system. At all stages, it was said to be provoked by plenitude and its signals were mediated by the nervous fluid.[5]

In *De ventriculo et intestinis*, Glisson promoted irritability to being a property that belongs to all the organs, or more specifically, to their component fibres. He visualized these fibres as extremely thin structures which serve as the basic functional elements of the muscles, tendons, nerves, and other organs. They were supposed to be round, flexible, resistant to tearing, extensible, and, of course, irritable. More than a century later, Bichat was to assign the vital properties, including irritability, to elemental tissues, which played much the same role in the body as Glisson's fibres. Both tissues and fibres were seen to be living units that function by responding to irritation or stimuli. Glisson's language, in fact, was adopted eagerly by such as Haller and the Montpellier physicians long before Bichat's time.

[2] Owsei Temkin, 'The classical roots of Glisson's doctrine of irritation', *Bull. Hist. Med.*, 1964, **38**: 297–328.

[3] Ibid. See also Walter Pagel, 'Harvey and Glisson on irritability with a note on Van Helmont', ibid., 1967, **41**: 497–514.

[4] In R. Milnes Walker, 'Francis Glisson', in Arthur Rook (editor), *Cambridge and its contribution to medicine*, London, Wellcome Institute, 1971, pp. 35–47. Though it is commonly accepted that Glisson was born in Dorset in 1597, Walker presents evidence that he was, in fact, born in Bristol in 1598 or 1599.

[5] Glisson's notion of irritability has been discussed in many places. Among them are Charles Daremberg, 'Glisson – Théorie de l'irritabilité et de la sensibilité', *Histoire des sciences médicales*, 2 vols., Paris, Baillière, 1870, vol. 2, pp. 650–672; Eyving Bastholm, 'Francis Glisson', *The history of muscle physiology*, trans. by W. E. Calvert, Copenhagen, Munksgaard, 1950, pp. 219–225; Walter Pagel, 'The reaction to Aristotle in seventeenth-century biological thought', in E. Ashworth Underwood (editor), *Science, medicine, and history*, 2 vols., Oxford University Press, 1953, vol. 1, pp. 489–509; Walker, op. cit., note 4 above, Pagel, op. cit., note 3 above.

Glisson distinguished three stages within the process of irritation. These are perception, the fibre's reception of an impulse; appetite, which awakens a desire on the part of the fibres to react; and the motion or execution of a required action. The sensation or perception of plenitude and the consequent irritation or contraction of bodily parts are, therefore, most intimately linked together.

Most bodily processes, of course, are automatic and unconscious, the brain and nervous system not being involved. Considering this unperceived local activity to be confined to the fibre, Glisson labelled it natural perception (*perceptio naturalis*), claiming that it presided over the activity of the heart, stomach, intestines, glands, and so on. He distinguished it from sensual perception (*perceptio sensitiva*), which involves the brain by way of the nerves, and from animal perception (*perceptio animalis*), which is conscious, deliberate, and under control of the will.[6]

As Pagel has shown, this *perceptio naturalis* owed much to the work of Helmont and William Harvey. To account for the development of an egg in his 1651 work *On generation*, for example, Harvey located a kind of irritability or motility in that egg itself, describing the property as "an animal virtue . . . with a principle of motion, of transformation, of rest and of conservation." Most interestingly, he went on to say that the nature of this "virtue" is such that "if one removes all obstacles from it, it will take the form of an animal."[7] Far from living activity being somehow imposed upon the body as the dualists would have it, Harvey assumed an embryonic plastic force which is responsible for formation of the body and which exists prior to the conscious soul and prior to any organs; that is, it belongs in the matter itself. Indeed, it is this very immanence of living forces within matter, at least certain kinds of matter, that is the important feature for us in the systems of Helmont, Harvey, and Glisson. These forces are manifest whenever a particular organization occurs.[8]

For monists then, a basic primeval level of life in the form of natural perception divorced from all sensation may be considered to belong to matter in general.[9] As we shall see, the basic theme of living forces being released from captivity by a particular form of organization seized the imagination of the *philosophe* Denis Diderot. In the meantime, however, as we have already seen, such views were eclipsed for some fifty years by the view that matter is inert and dependent upon external forces for its motion. Accordingly, iatromechanism and animism divorced matter from its life, thereby separating elements that many persons had hitherto considered to be indissolubly linked together in biological units.

The language of irritability of living parts was resurrected in the 1740s. Bordeu and Haller both built their physiological systems around it. Bordeu fused it on to animists' arguments against iatromechanism. It was a fruitful amalgam which gave him the means for turning his back upon the increasingly limiting dualist systems. Bordeu's work on glandular function easily owed at least as much to that of the seventeenth-

[6] Francis Glisson, *Tractatus de ventriculo et intestinis*, London, E. F. for H. Brome, 1672, pp. 147–174.

[7] William Harvey, *Exercitationes de generatione animalium*, London, Typis Du Gardianis, 1651, p. 77. Quoted by Jacques Roger, *Les sciences de la vie dans la pensée française du XVIIIe siècle*, Paris, Colin, 1971, p. 114.

[8] Pagel, op. cit., note 3 above; Temkin, op. cit., note 2 above.

[9] A point made by Pagel, op. cit., note 5 above.

century monists as it did to Stahl and Sauvages.[10] The force of sensibility, as he conceived of it, had much in common with both the *archeus*, and with Glisson's natural perception. As early as 1742, Bordeu had argued in his bachelor's thesis that there is a sensibility proper to each organ and that the soul or vital force resides in the nerves. In his doctoral thesis one year later, he speculated that the stomach contracts because its membrane possesses some critical force. Here in embryo were the major themes which evolved into his mature system of physiology.

Bordeu saw irritability and sensibility as being so closely linked as to be inseparable. In the case of salivary glands, for example, he claimed that the activities of eating and speaking provided the irritation which awakens sensibility, which in turn provokes the secretion of humour. The following quotation from the *Recherches sur le pouls* demonstrates that he conflated the two properties, insisting that nerves are essential for sensibility to exist.

> Each organic part of the living body has nerves which have a *sensibility* . . . it is the necessary result of their constitution, of their position and their modification in the body or in its parts, when they are not entirely deprived of the conditions in which life can neither be demonstrated nor exist; *sensibility* is of different types, and is more or less co-mingled with *mobility* or *contractility* . . . . All the ancient philosophers and physicians thought pretty much the same . . . the *strictum* of the systematists, the tonic motion, fibrillatory motion, stimulus, irritation, setting on edge of nerves, spasm, modern contractility, all this is explained by pretty much the same idea; this is by the *activity* of nerves, the scope of this *activity*, a virtue, a property, a particular disposition which Glisson called *irritability*.[11]

Nerves transport critical signals which activate the organs and they integrate the actions of the various organs and parts. In the case of glands, for example, they signal the need for a particular humour: they accompany the arteries controlling the blood flow in the region of the gland; and they carry the irritating signal which initiates the secretory process directing glandular sphincters to accept or reject the particular parts of the blood presented to them. While nerves act as pathways for the transmission of directions it is each organ's particular indwelling sensibility that defines its nature and activity. "The eye cannot endure oil which the stomach retains easily and the latter rejects the emetic which makes practically no impression upon the eye." Each part reacts in an active manner to its own peculiar sensibility. "The eye prepares itself to receive light . . . . The ear strains itself, opens itself, finally disposes itself to receive sound waves."[12]

The best known of all work on the subject of sensibility and irritability, however, is that of *Albrecht von Haller* (1708–77). A pious and peevish man beset by self-doubts and haunted by religious insecurities, he had a special knack of making professional enemies. Born in Berne, he went to Leiden to study medicine under Boerhaave. Invited in 1736 to be Professor of Medicine at the newly founded University of Göttingen, he remained there for seventeen productive but largely unhappy years. In 1747, Haller published the first edition of his *Primae lineae physiologiae*, which has been described as Europe's first medical textbook. In 1753, much to the annoyance of Göttingen, he

[10] See also Elizabeth L. Haigh, 'Vitalism, the soul and sensibility; the physiology of Théophile de Bordeu', *J. Hist. Med.*, 1976, **31**: 30–41.

[11] Théophile de Bordeu, 'Recherches sur le pouls', in *Oeuvres complètes*, 2 vols., Paris, Caille et Ravier, 1818, vol. 1, pp. 420–421. Quoted by Roger, op. cit., note 7 above, p. 626.

[12] Bordeu, 'Recherches anatomiques sur la position des glandes et sur leur action', in *Oeuvres complètes*, op. cit., note 11 above, vol. 1, pp. 156–165.

gave up his teaching career and medical practice to return to Berne, where he became a court bailiff and, in 1758, a director of a salt works. While a civil servant, he published an eight-volume *Elementa physiologiae.* His decision to give up medical practice in spite of his considerable reputation probably had to do with a revulsion against vivisection at a time when most of his fellow experimenters appear to have been indifferent to the bitter price exacted from animals. In any case, Haller does not seem to have found peace of mind, for he died an opium addict.[13]

Haller confused the discussion of sensibility and irritability of the bodily parts somewhat by insisting that the two terms ought to be defined differently from the way in which Glisson and Bordeu had used them. He equated sensibility with conscious feeling and irritability with observable motion. Upon examining the distribution of the properties, he found that irritability is largely a property of the muscular fibre and sensibility is basically limited to nerves. Furthermore, he observed that the more sensible organs generally possess relatively little irritability while the more irritable ones have little sensibility.

Haller also discussed an automatic and unconscious tendency to self-motion which belongs to muscle fibres. He ascribed to it properties and functions that were much like those Bordeu had assigned to sensibility. *First lines*, in which Haller first described this reactivity, opens with an important discussion of animal fibres, which Haller believed to be the structural elements of the body. "A *fibre*, in general", he wrote, "may be considered as resembling a line made of points, having a moderate breadth; or rather as a slender cylinder." He discerned two types of fibres. One is longitudinal and occurs in bones, tendons, ligaments, and muscles, while the other is flatter and occurs in cellular membranes and vessels.[14] Boerhaave had had a similar concept of multitudinous small vessels serving as the basis of all the organs and structures,[15] and it is possible that he and Haller were finally indebted to Glisson, who had discussed fibres before either of them. On the other hand, the general notion of fibres as smallest structures may equally well have been but the inevitable adaptation of the corpuscular philosophy that had emerged earlier in the physical sciences. In any case, the concept permitted Haller to adapt a modified version of Glisson's irritability and Bordeu's sensibility to the physiology that he inherited from Boerhaave.

On the heart's apparently innate ability to move, Haller declared:

> There resides in the heart a kind of impatience to stimulus .... This irritability is greater and remains longer in the heart than in any other part of the body; seeing, by stimulating it, the motion of the heart may be renewed at a time, when that of no other muscle can .... That motion is *peculiar to the heart itself*; coming neither from the brain, nor the soul; seeing it remains in a dead animal even when the heart is torn out of the breast; neither can it, by any act of the will, be made either quicker or slower.[16]

He labelled this "impatience to stimulus", this "irritability" which derives from the fibres themselves, a *vis insita* (resident force). It is particularly abundant in the heart

[13] An excellent and perceptive account of Haller's theory and of its central role in the development of physiology is that of François Duchesneau, *La physiologie des lumières*, The Hague, Nijhoff, 1982, pp. 141–234. See also Lester S. King, 'Introduction', to Albrecht von Haller, *First lines of physiology*, trans. and ed. by William Cullen, Göttingen, Wrisber, 1786, reprinted New York and London, Johnson Reprint Corp., 1966, vol. 1, pp. xii-xxiii.

[14] Haller, op. cit., note 13 above, vol. 1, sect. XV, pp. 9–15.

[15] King, 'Introduction', op. cit., note 13 above, vol. 1, pp. xxiii—xxxii.

[16] Haller, op. cit., note 13 above, vol. 1, sect. C11, pp. 59–60. The emphasis in the passage is Haller's own

and intestines, which are observed to contract long after they are removed from the body. It is activated by different stimuli in different parts with the result that the bladder responds to urine, the heart to blood, and so on.[17] He very clearly distinguished this particular reactivity from that which derives from the will:

> The heart and intestines, also the organs of generation, are governed by a vis insita, and by stimuli. These powers do not arise from the will; nor are they lessened or excited, or suppressed, or changed by the same. No custom no art can make these organs subject to the will, which have their motions from a vis insita; nor can it be brought about, that they should obey the commands of the soul, like attendants on voluntary motion. It is so certain that motion is produced by the body alone, that we cannot even suspect any motion to arise from a spiritual cause, besides that which we see is occasioned by the will, as stimulus will occasion the greatest exertions, when the mind is very unwilling.[18]

Haller described yet another active but automatic force which belongs to animal and vegetable fibres, including even the hair, feathers, membranes, and the cellular texture. It is a very slow "contractile power", which resists the stretching of the fibres and restores them to their normal size when the extending power is removed. It is observed in cases of pregnancy, obesity, and so on. But they are dead forces, he said, efficacious after death and not to be confused with such living forces as the *vis insita*, sensibility, and irritability.[19] His description of this force was much like that which Barthez subsequently labelled the *force de situation fixe* and described as a resistance to attempts to disrupt the organism.[20] Later still, Bichat stated that a slow extensibility and contractility are properties not of life but of texture.[21]

The muscular power which derives from the will or the soul alone depends upon the nerves. "For the nerve alone has feeling; this alone carries the commands of the soul; and of these commands there is neither instinction nor perception in that part, whose nerve is either tied or cut, or which has no nerve."[22] Whereas the *vis insita* remains after a nerve is cut, the willed power depends absolutely upon an intact nervous supply. This willed contractility and the nervous sensibility were the subjects of two papers which Haller read before the Royal Society of Göttingen in 1752. Entitled 'On sensibility' and 'On irritability', they contributed much to Haller's considerable reputation. They were published in many editions, each successive edition growing longer with supplements largely added in response to various critics who took exceptions to Haller's conclusions. Simon Tissot, a Swiss physician and a friend of Haller, published a French translation of the work in 1755. That same year, there appeared an anonymous English translation based upon Tissot's and even using his preface.

In it, Haller made it clear at the very beginning that his concern was with conscious and observable properties:

> I call that part of the body irritable which becomes shorter upon being touched; very irritable if it contracts upon a slight touch, and the contrary if by a violent touch it contracts but little. I call that a

[17] Ibid., vol. 1, sects. CCCCII–CCCCIII, pp. 233–234.

[18] Albrecht von Haller, 'A dissertation on the sensible and irritable parts of animals' (London, J. Nourse, 1755), a contemporary translation from Latin with introduction by Owsei Temkin, *Bull. Hist. Med.*, 1936, **4**: 651–699.

[19] Haller, op. cit., note 13 above, vol. 1, sects. CCCXCI-CCCXCII, pp. 226–227.

[20] Paul Joseph Barthez, *Nouveaux éléments de la science de l'homme*, Montpellier, J. Martel ainé, 1778, pp. 68–82.

[21] Bichat, op. cit., note 1 above, p. 130.

[22] Haller, op. cit., note 13 above, vol. 1, sect. CCCCIV, p. 235.

> sensible part of the body, which upon being touched transmits the impression of it to the soul; and in brutes, in whom the existence of the soul is not so clear, I call those parts sensible, the irritation of which occasions evident signs of *pain* and disquiet in the animal. On the contrary, I call that insensible, which being burnt, tore, pricked, or cut till it is destroyed, occasions no sign of pain nor convulsion, nor any sort of change in the situation of the body.[23]

To determine the distribution of sensibility and irritability, Haller dissected animals of various kinds and ages. After exposing a particular part, he waited until the animal ceased to struggle and complain. To test for sensibility, he subjected the part in question to a blow, to heat, alcohol, the scalpel, acids, and so on. If the animal lost its tranquillity, began to complain, or withdrew the wounded part, Haller concluded that it had sensibility. He found that the skin is very sensitive "for in whatever manner you irritate it, the animal makes a noise, struggles, and gives all the marks of pain which it is capable of." By the time he dissected down to the fat and cellular membrane, sensibility appeared to have vanished. Deeper still he found that muscular tissue responds to painful stimuli but only if its nerve supply is intact. Tendons are absolutely insensitive, for an animal in whom such an organ is lacerated, burned, or pricked shows no sign of pain. When it is released, provided that even a small part of the tendon remains intact, the animal walks easily apparently with no pain.

In *First lines*, Haller had asserted that nerves alone carry sensibility. He stuck to that conviction in this later work but by so doing, he was forced into certain inconsistencies. Teeth have a nerve supply and as expected, he found them to be sensitive. By the time he came to examine bones, the animal tested was in such pain that no accurate observations were possible. Though bones are made of the same basic substance as teeth, Haller argued, they must be insensible since they have no discernible nerve supply. Though other people had frequently observed that marrow is very painful, Haller thought that to be improbable "as it is of the same nature with fat, and has no nerves bestowed upon it." When faced with the difficulty of explaining why such viscera as lungs, liver, spleen, and kidneys are sensitive though they possess no nerve supply, Haller pleaded rather weakly that "I do not allege that they are void of all sensation, but only that it is very weak in them, *viz.* such as one would expect in a part which has very few nerves bestowed on it in proportion to its bulk."[24]

Haller found that irritability and sensibility exist in a kind of inverse relationship in the parts. While irritability may be mediated through a nerve, there is never any motion in the nerve itself. While cutting or tying the nerve to a part destroys its sensibility, it does not affect contractility. That is, parts of the body that are normally subject to the will are largely paralysed if a nerve is cut, but they retain an ability to contract in response to direct stimulation.[25]

[23] Haller, op. cit., note 18 above, pp. 658–659.

[24] Ibid., pp. 658–673.

[25] Haller found that the *sensible organs* of the body are the skin, muscles, teeth, retina, stomach and intestines, genitalia, heart, and nerves. Only a little sensibility is present in the glands and in internal viscera such as the liver, spleen, kidneys, and lungs. The *insensible organs* are fat, cellular membranes, tendons, periosteum, bones, marrow, dura mater and pia mater, omentum and pleura, blood vessels, cornea and iris, and capsulae of the joints. The *irritable organs* are the muscles (including the heart and diaphragm), stomach and intestines, caecum, glands, genitalia, bladder, and lacteals. The *non-irritable organs* are the skin, tendons, periosteum, dura mater and pia mater, mediastinum, omentum and pleura, blood vessels, pericardium, and nerves. This information and the description of the experimentation which led to the conclusions is presented in op. cit., note 18 above.

This work was contemporary with Bordeu's examination of the glands. Haller attempted to discredit this system of "M. *Du Bordeu*, so severe a critic of the writings of others" when he had occasion to comment upon it. According to Haller, the glands in general have very little sensation for they are supplied with very few nerves. He wrote that "there are no considerable nerves besto'd upon the largest of the glands, nor the *thymus*", and "the nerves which go to the thyroid gland are a great deal smaller than those of a muscle ten times less in bulk than that gland."[26] I suspect that Haller was succumbing to a touch of petulance if not downright jealousy, for Bordeu's work was unquestionably impressive. Curiously enough, Boerhaave himself had remarked that glands "receive a great many Nerves, more in Proportion to their Bulk, than any other Body; and which are distributed so minutely throughout the whole Body of the Gland, that they seem to occupy every individual Point."[27] It may be that Haller disliked Bordeu's conclusions because they derived from a vitalist orientation. In any case, he disregarded his impeccably made observations and affirmed his own faith in the mechanist viewpoint, which held that pressure upon a gland causes it to expel its juices: saliva, for example, is squeezed out in the mouth when we are not hungry by means of the compression of the digastric muscle.[28]

Bordeu responded, albeit mildly, to Haller's criticism in his *Recherches sur l'histoire de la médecine*. Though he remarked that Haller's theory was but one of a number on the subject of irritability, he described Haller as "one of the most distinguished medical philosophers of this century". But he pointed out that the Montpellier school had taken the irritability of the parts of the living body to be a general principle before it was studied from Haller's particular viewpoint.[29]

Haller's most important and long-lasting argument surrounding the question of the sensible and irritable parts was with *Robert Whytt* (1714–64), who had studied medicine in London, Paris, and Leiden before receiving a medical doctorate from Rheims in 1736. In 1746, he took an appointment as professor of medical theory in his home town of Edinburgh, where he quickly developed a reputation as a major physician. He became a fellow of the Royal Society, the first physician to the king in Scotland and in 1763, president of the Royal College of Physicians of Edinburgh.

Like many of his contemporaries, Whytt was interested in the larger subject of organic motion. In 1751, he published an *Essay on the vital and other involuntary motions of animals*, a book which is judged to be his single most important work.[30] Whytt's ideas are particularly interesting insofar as he applied animist notions of life and matter to the study of animal activity. Though he had gone to Leiden specifically to hear Boerhaave, Whytt was not only not persuaded by mechanism but downright disdainful of anyone who was. He described the idea that an inanimate machine can

[26] Ibid., p. 67.

[27] Herman Boerhaave, 'Of the different structure of the glands', *Dr. Boerhaave's academical lectures on the theory of physic, being a genuine translation of his Institutes as dictated to his students at the University of Leyden*, 6 vols., London, W. Innys, 1757–73, vol. 2, sect. 242, pp. 210–211.

[28] Haller, op. cit., note 18 above, pp. 673–678.

[29] Bordeu, 'Recherches sur l'histoire de la médecine', in *Oeuvres complètes*, op. cit., note 11 above, vol. 2, pp. 668–669.

[30] Whytt's *An essay on the vital and other involuntary motions of animals* is discussed at considerable length by Roger French in 'Involuntary motions and the reflex', *Robert Whytt, the soul, and medicine*, London, Wellcome Institute, 1969, pp. 79–92.

produce heart movement, respiration, or intestinal motion simply by virtue of its mechanical construction as "a notion of the animal form too low and absurd to be embraced by any but the most minute philosophers!"[31] He added that since the contraction of the heart, motion of the blood, and continuance of life in general cannot be accounted for mechanically, one must have recourse to the "energy of a living principle capable of generating motion".[32]

Whytt dismissed the monists as decisively as he had the mechanists, for he simply could not admit the possibility that animal fibres possess a power of sensation and of generating motion unless these are united by "an active PRINCIPLE, as the SUBJECT and CAUSE of these".[33] He considered the soul to be the living and unifying principle, insisting, however, that he was not a Stahlian. He particularly objected to Stahl's contention that the soul acts rationally when it is directing unconscious activity.[34] He preferred to consider it to be a sentient rather than a rational principle:

> The mind . . . in producing the vital and other involuntary motions, does not act as a rational, but as a sentient principle; which, without reasoning upon the matter, is as necessarily determined by an ungrateful sensation or *stimulus* affecting the organs, to exert its power, in bringing about these motions, as is a balance.[35]

It is curious that while Whytt rejected the notion of sentient matter, he felt comfortable in imposing a sentient principle, that is the soul, to move the inert matter. It is a good illustration of the tenacity of the mind-matter dualism so characteristic of the seventeenth century. In the *Essay*, Whytt wrote that irritability is a power of the soul, which is awakened by the sentient activity that resides in the brain and the nerves. The muscle which is about to contract "perceives the stimulus". Thus, as with Bordeu, irritability depends upon sensibility and is its inseparable adjunct. In turn, that sensibility is a function of the central nervous system.

Haller reviewed Whytt's *Essay*, dismissing what he called the specious Stahlian assumption that all motions proceed from the soul. Unconcerned about the distinctions separating animists, Haller continued thereafter to group Whytt with the Stahlians, apparently oblivious to any number of protests to the contrary.[36] Whytt, in turn, reviewed Haller's work on sensibility and irritability in a 1755 treatise entitled *Observations on the sensibility and irritability of the parts of man and other animals*. He acknowledged in the introduction that these "new and curious experiments" must henceforth produce considerable changes in the theory and practice of medicine. He also took the opportunity to strike back at Haller's criticism of him by caustically observing that Haller "has taken uncommon pains in making many and repeated experiments; as much to overpower the incredulous by their number as to secure himself from any chance of being deceived." He proceeded to challenge many of

[31] Robert Whytt, *An essay on the vital and other involuntary motions of animals*, Edinburgh, Hamilton, Balfour & Neill, 1751, p. 2.

[32] Ibid., pp. 270–271.

[33] Ibid., pp. 241–242.

[34] Ibid., p. 267.

[35] Ibid., p. 289.

[36] French, 'The soul in physiology', in op. cit., note 30 above, pp. 149–160. French discusses the development of the notion of the physiological role in general as well as Whytt's particular conception of the soul's activity.

Haller's observations and conclusions concerning the reactivity of the parts. Haller responded to these observations in the 1756 edition of *On the sensible and irritable parts of animals*, to which Whytt replied again in a 1768 edition of *Observations*.[37]

Whytt's major criticism of Haller's experimental technique was an elementary one, for he simply pointed out that a great pain will dampen or destroy a lesser one. By the time Haller had opened an animal's thorax, he wrote, the beast was suffering excruciating pain so that the fact that it did not register additional pain when its heart was pricked or cut did not justify the conclusion that the heart is without feeling.[38] In general, Whytt challenged many of Haller's basic observations. He refuted Haller's claim that the periosteum and the tendons are insensitive by pointing out that arthritis is very painful. Haller replied that the seat of arthritic pain is in the skin or the subcutaneous nerves. Neither, he added, does phrenitis occur in the dura mater or pleurisy in the pleura, as is commonly assumed, for his own experiments had shown that the dura mater and the pleura are devoid of feeling. Though many physicians report that pain is experienced in the bone marrow, Haller insisted that the organ is insensitive because it is "cellular", of the same nature as fat, and with no visible nerves. When Whytt pointed out that many persons had been heard to complain that the marrow of a cut arm or leg was painful to touch, Haller replied that a single experiment was not sufficient to prove marrow sensibility.[39]

Haller listed the cornea, kidneys, and glands among the insensitive organs. Whytt pointed out that one feels one's cornea when it is pressed, that a patient feels pain when a kidney is opened in surgery, and that a blow to the testicles or breasts produces much pain. He wrote appropriately that "One instance of this kind is more decisive . . . than twenty experiments on brutes who cannot inform us whether they feel a slight pain or none at all."[40] Thus the debate proceeded with neither man acknowledging the legitimacy of the other's observations. Whytt denied the existence of an altogether insensitive organ or tissue, pointing out that even those parts that normally seem to be so become painful if they are diseased or inflamed. Though Haller never conceded much credibility to Whytt, it was largely due to peevishness. Bichat was to take up Whytt's point by stating that a normally unconscious sensitivity becomes conscious whenever it is augmented by disease or inflammation.

However much Haller might have objected to having his notions challenged or developed by vitalists like Bordeu or Whytt, the trauma was as nothing compared to that of having them linked with the ideas of *Julien Offray de La Mettrie* (1709–51). The deeply pious Haller was appalled and offended by the work of this outspoken

[37] This is described in considerable detail by French, 'The controversy with Haller: sense and sensibility', ibid., pp. 63–76. French also discusses it in 'Sauvages, Whytt and the motion of the heart: aspects of eighteenth-century animism', *Clio medica*, 1972, **7**: 35–54. The metaphysical and theoretical differences between the two men are perceptively discussed by Duchesneau, op. cit., note 13 above, pp. 171–234.

[38] Robert Whytt, 'Observations on the sensibility and irritability of the parts of man and other animals', in *The works of Robert Whytt*, Edinburgh, T. Becker & P. A. de Hondt, 1768, pp. 251–261.

[39] Haller, op. cit., note 18 above, p. 667.

[40] Haller's responses to Whytt's 'Observations on sensibility and irritability' (op. cit., note 38 above) are in his *Mémoires sur la nature sensible et irritable des parties du corps animal*, trans, by Simon André Tissot, 4 vols., Lausanne, Marc. Mic. Bousquet, 1756, vol. 4, pp. 102–134. In a 1768 edition of his work, Whytt responded in turn to Haller's responses and thus the argument continued. This is discussed by French in 'The controversy with Haller', op. cit., note 37 above pp. 68–76.

materialist, who was widely vilified as a libertine and an atheist. La Mettrie used basic organicist views and the notions of sensibility and irritability to develop the thesis that psychic or rational phenomena in man depend not upon an immaterial soul but upon physical factors alone. By so doing, he also denied the spiritual element of the human being.

La Mettrie was a native of Brittany.[41] He attended the Paris Medical School for five years but, like many students at the time, he transferred to Rheims to get his medical doctorate at less cost. Then he went to Leiden to study with Boerhaave. He undertook to translate and to annotate the famous physician's medical work, bringing out an abridged translation of Boerhaave's *Institutiones rei medicae* shortly after Haller had completed an edition of it in 1743. To explain muscular contraction, Boerhaave had relied upon the traditional albeit clumsy notion of animal spirits flowing into the muscles. Perceiving the limitations of that theory, Haller had looked for other explanations of muscular motion, as we have seen. La Mettrie was familiar with the work of Haller and very probably with that of other contemporaries, who had examined the nature of muscular activity and of irritability.[42] The evidence of an organic reactivity became the theoretical tool by means of which La Mettrie gave coherence to his thesis that life can exist apart from any sort of a soul or other vital principle.

In 1742, La Mettrie abandoned his family and a medical practice in his native Saint Malo to go to Paris, where he met many physicians and notable *philosophes*. While serving as an inspector of hospitals for the armies on campaign, La Mettrie composed his *Histoire naturelle de l'âme*, wherein he argued that thought, volition, and all purposive motions of the body are the products merely of the physical organization of the body, a necessary consequence of the unique arrangement of its parts. A decree of the Parlement of Paris in July 1746 acknowledged the importance of the work by condemning it to the flames. La Mettrie left a hostile France for the more liberal Holland.

The following year, his most famous work, *L'homme machine* was published. It appalled even the Dutch. The abuse heaped on La Mettrie reached hysterical proportions, forcing him to take refuge in the court of Frederick the Great in Berlin. The suave and worldly emperor's repulsion at his own pious Lutheran upbringing caused him to be tolerant of persons whom he saw to be victims of bigotry. Writing about La Mettrie, he observed that "Calvinists, Catholics and Lutherans forget . . . that co-substantiation, free will, the mass for the dead and the infallibility of the pope divide them." Like almost everything that emerged from La Mettrie's pen, *L'homme machine* delivered jibes at those persons he thought were too pompous, pious, or self-righteous. The work was pointedly antagonistic to established religion, and even atheistic in its viewpoint; it dismissed the soul as "merely a vain term about which no one has any idea",[43] and life after death as "a chimera based upon absurd

[41] Julien Offray de La Mettrie, *L'homme machine*, edited by Aram Vartanian, Princeton University Press, 1960: La Mettrie's life and work are described by Vartanian in his 'Introduction', pp. 1–12. A brief survey of La Metttrie's biological ideas is found in Thomas S. Hall, *Ideas of life and matter*, 2 vols., University of Chicago Press, 1969, vol. 2, pp. 46–56.

[42] Vartanian, op. cit., note 41 above, pp. 84–89, discusses La Mettrie's debt to Haller.

[43] Ibid., p. 180.

reasoning".[44] The European scholarly world recoiled in indignation.[45]

Today, the thesis of *L'homme machine* does not strike us as being any more fundamentally atheistic than that of any other physiological work described herein. The crucial point is that La Mettrie moved a logical and inevitable step beyond Bordeu, Whytt, and Haller by extrapolating organicism to its logical monist conclusion, maintaining that the rational and conscious processes are a part of the matter of the body. If all living processes derive from organization, there remains no need for an immaterial overseer. The assertion quoted above concerning the "vain term" occurred in the following context:

> All the faculties of the soul depend on the proper organization of the brain and of the body so that they are visibly nothing but organization . . . the soul, therefore, is merely a vain term about which no one has any idea and for which a good intellect can only serve to name the part of us which thinks. Given the least principle of motion, animated bodies will have everything necessary for them to move, to sense, to think, to be contrite, and in a word, to conduct themselves in the physical realm and in the moral one which depends upon it.[46]

La Mettrie offered a number of relevant but common observations to demonstrate the existence of a motive principle in living flesh. He remarked that all animal muscle palpitates after death; muscles separated from the body contract when they are pricked or otherwise stimulated; intestines retain their peristaltic motion for a considerable time after death or outside the body; a simple injection of warm water will reactivate the heart or muscles and so on. He was fascinated by a remarkable little creature called a polyp, which, superficially at least, resembles a plant more closely than an animal. A naturalist, Abraham Trembley, had observed that any part separated from the creature can, under certain circumstances, regenerate a complete new polyp in a few hours. It had come to be a kind of *cause célèbre* of the scientific world and its regenerative capacity lent itself to a variety of interpretations. Whytt, for example, believed that it demonstrated the presence of a ubiquitous soul in living matter, while Barthez was confirmed in his theory of the vital principle. For La Mettrie and for monists in general, the polyp's fascinating ability was powerful evidence for a motive force belonging to the parts themselves. It could not help but confirm La Mettrie in his conviction that life has nothing to do with a vital or spiritual principle.[47]

If the source of intelligence and personality is but a consequence of the arrangement of the parts, it follows that the differences between man and the animals are due merely to the degree or quality of their respective organizations. It is conceivable, La Mettrie wrote, that education might be able to bridge the gap between man and the higher animals. In the *Discourse on method*, Descartes had emphasized the uniqueness of human speech, claiming that it is an ability conferred by the soul. La Mettrie, however, speculated that apes might well be capable of learning to speak.[48] Although some recent anthropologists have come to believe that such speculations may have some foundation, in the middle of the eighteenth century, the suggestion of

[44] Ibid., p. 196.

[45] Ibid., p. 95–113. Vartanian describes the reception of *L'homme machine* here.

[46] Ibid., p. 180.

[47] Ibid., pp. 180–182. See also Aram Vartanian,'Trembley's polyp, La Mettrie, and eighteenth-century French materialism', *J. Hist. Ideas*, 1950, **11**: 259–286.

[48] La Mettrie, op. cit., note 41 above, pp. 161–162.

such a possibility was deemed outrageous, which may be precisely why La Mettrie wrote it down.

La Mettrie's man-machine has often been interpreted as an extrapolation of Descartes' beast-machine. King, for example, wrote that "While Descartes had regarded animals as machines, the dualistic philosophy gave to man a soul which the animals lacked and which differentiated a human from a machine. It was only a small step, but a mightily important one, to say that the mind of man was not a separate substance."[49] Indeed, the title which La Mettrie chose for his work and much of his language in the text invites that interpretation. It may even be that the notion of the beast-machine sparked the initial idea which finally produced the work. Nevertheless, a large conceptual gulf separates the two men, for La Mettrie utterly rejected dualism. The machine he described is composed not of inert, brute matter but of material substance which is vibrantly alive and throbbing with activity. We read, for example, that "The human body is a machine which winds its own springs. It is the living image of perpetual movement."[50] And, "Let us enter into some detail concerning the innate activity of the human machine. All the vital, animal, natural and automatic notions are due to its action."[51] Near the end of the work, he affirmed, "Let us conclude thus boldly that man is a machine, and that in the whole universe there is but a single substance differently modified."[52] Therefore, though man is a machine, he is no mere mechanism. Much as La Mettrie must have been imbued by Boerhaave with the notion of mechanical laws governing an organism's behaviour, he was also much indebted to investigations of the motive principles resident in matter itself.[53]

Though La Mettrie's aggressive atheism appalled his contemporaries, they could not help but recognize him as a prodigal intellectual relation. He maliciously pointed out the theoretical kinship that linked their work to his own. Jerome Gaub, Boerhaave's successor at the University of Leiden, is a case in point. A few months before *L'homme machine* was completed, La Mettrie had heard Gaub read a paper, *De regimine mentis*, which concerned itself with the relationship between mind and body. Proceeding from the assumption that mental phenomena are physically regulated, Gaub proposed the possibility of thus understanding and hence controlling psychic and mental phenomena. Many of his examples were then borrowed by La Mettrie, to the consequent chagrin of the good Christian, sorely tried for a long time by the apparent affinity between his notions and those of La Mettrie. In a 1763 paper, he referred to La Mettrie's presence at the talk he delivered sixteen years earlier:

> I do indeed regret bitterly that a little Frenchman ... brought forth a repulsive offspring, to wit, his mechanical man, not long after sitting before this chair and hearing me speak, and did this in such a way

[49] King, op. cit., note 13 above, p. xlii.

[50] La Mettrie, op. cit., note 41 above, p. 154.

[51] Ibid., pp. 182–183.

[52] Ibid., p. 197.

[53] Aram Vartanian has contributed substantially to our understanding of La Mettrie's place among eighteenth-century theorists. In his introduction to a recent critical edition of *L'homme machine* (op. cit., note 41 above, p. 36), he examined La Mettrie's intellectual links with both mechanists and monists, pointing out that his theory was not rooted in any traditional notion of an externally directed machine. The irritability principle led him to a concept of the human machine as a self-sufficient dynamic system of interdependent parts. He labelled La Mettrie's viewpoint a "vitalo-mechanist orientation", for, far from being an extrapolation of the beast-machine, the man-machine's substance throbs with indwelling vitality.

that it seemed to many people that I had furnished him with, if not sparks for his flame, at least matter for embellishing his monstrosity.[54]

The most famous conflict of all was with Haller. In his abridged translation of Boerhaave's *Institutes of medicine*, La Mettrie had borrowed freely from Haller's edition of the same work, but without acknowledging his debt. Haller charged him with deception in a review of his *Histoire naturelle de l'âme*. As if to compensate for his earlier neglect, the still obscure La Mettrie dedicated *L'homme machine* to Haller, prefacing the work with an effusive tribute from "your disciple and your friend". The dedication remained in the 1751 edition, in which La Mettrie also announced that Haller was his inspiration and his teacher.[55] The testy protestant was left now complaining that he wished no connexion with this "impious system which my experiments totally refute".[56] The point is, significantly enough, that his experiments by no means refuted La Mettrie, who was correct in seeing a relationship between their basic notions. It was, nevertheless, a cruel game, for the prim Haller was utterly unable to match wits with La Mettrie.

La Mettrie died at the age of fifty-two, reportedly due to an excess of a truffle pâté. His detractors interpreted his untimely end as an act of divine vengeance as if God himself had intervened to refute the materialist philosophy. Today, La Mettrie's well-placed jibes against the pompous and the pious make delightful reading. We are by now sufficiently removed from his milieu and the passions it engendered to see that he was correct to claim a relationship between his own work and that of Glisson and of many contemporaries, some of whom, like *Denis Diderot* (1713–84), even shared many of his ideas about the soul and matter, and agreed with his demonstration.

Neither a physician nor a naturalist himself, Diderot was much preoccupied with

[54] L.J. Rather, *Mind and body in eighteenth-century medicine. A study based on Jerome Gaub's De regimine mentis*, London, Wellcome Institute, 1965, pp. 115–204. Gaub's reaction to La Mettrie's work is also described by Vartanian, op. cit., note 41 above, pp. 90–92. Vartanian quotes from a letter which Gaub wrote to Charles Bonnet concerning that same 1747 lecture and La Mettrie's attendance at it. He said: "Des esprits malins en tirent des conséquences irreligieuses."

[55] La Mettrie dedicated *L'homme machine* to Haller with the following words: "C'est le plaisir que j'ai eu à composer cet ouvrage, done je veux parler; c'est moi-même, & non mon livre que je vous addresse, pour m'éclairer sur la nature de cette sublime Volupté de l'Etude. Tel est le sujet de ce Discours. Je ne serois pas le premier Écrivain qui, n'aient rien à dire, pour réparer le Stérilité de son Imagination, auroit pris un texte où il n'y en eut jamais. Dites-moi donc, Double Enfant d'Apollon, Suisse Illustre, Fracastor Moderne, vous qui savez tout à la fois connoître, mesurer la Nature, qui plus est la sentir, qui plus est encore i'exprimer; savant Médecin, encore plus grand Poëte, dîtes-moi par quels charmes l'Étude peut changer les Heures en moments; quelle est la Nature de ces plaisirs de l'Esprit, si différents des plaisirs vulgaires .... Mais la lecture de vos charmantes Poësies m'en a trop pénétré moi-même, pour que je n'essaie pas de dire ce qu'elles m'ont inspiré. L'homme, consideré dans ce point de vue, n'a rien d'étranger à mon sujet." Op. cit., note 41 above, p. 143.

[56] Haller voiced his personal objections to *L'homme machine* as follows: "The deceased M. de la Mettrie has made Irritability the basis of the system which he advanced against the spirituality of the soul; and after saying that Stahl and Boerhaave knew nothing of it, he has the modesty to assume the invention to himself, without ever having made the least experiment about it. But I am certainly informed that he learnt all he knew about it from a young Swiss with whom I am not acquainted: who never was my pupil, not is he a physician, but he has read my works, and seen some of the famous Albinus's experiments and upon these La Mettrie built his impious system, which my experiments totally refute. For if Irritability subsists in parts separate from the body and not subject to the command of the soul, if it resides every where in the muscular fibres and is independent of the nerves, which are the *satellites* of the soul, it is evident, that it has nothing in common with the soul, and it is absolutely different from it; in a word, that neither Irritability depends upon the soul, nor is the soul what we call Irritability in the body." Op. cit., note 18 above, pp. 695–696.

scientific and technological development. Indeed, he acquired a considerable proficiency in mathematics, physics, chemistry, physiology, and a number of languages. Acutely aware of the great changes which science and the modes of thought accompanying it were producing upon the intellectual milieu of Europe, he plunged boldly and with considerable insight into the examination of many suggestive ideas and their implications. Among the questions that interested him most were those having to do with life, consciousness, and rationality.

Diderot's best-known accomplishment is the popular and vastly influential thirty-five-volume *Encyclopédie* completed in 1780, after some twenty years of toil, but in addition, he wrote extensively on a variety of subjects.[57] For our purposes, it is instructive to examine the development of Diderot's ideas about living processes. His mature theory, developed by 1769, was so provocative that he dared not publish it. By then, he had come to believe that the whole universe is composed of one single substance which possesses sensitivity. Life, he argued much like Harvey, is not the imposition of vital forces on to matter but a release of sensitivity from its material prison.

In 1746, Diderot composed *Pensées philosophiques*, a work preoccupied with nature's beauty, which he took to be a witness to divinity, but not to the malevolent Christian God whom he eschewed. His more benevolent deity was a kind of master mechanic who oversees natural phenomena including organic processes. The Parlement of Paris found the book to be unacceptable for general reading and condemned it to be burned.[58]

By 1749, when his *Lettre sur les aveugles à l'usage de ceux qui voient* appeared, Diderot had effectively evolved into an atheist. Relating therein a fictitious conversation between a dying mathematician, Nicholas Saunderson, blind from birth, and a Reverend Holmes, Diderot has Holmes trying to persuade Saunderson that one can discern the existence of God in the complex mechanism of the organs, and in the beauty and order of nature generally. Saunderson protests that this need has nothing to do with a sovereignly intelligent being. "If it is a matter of astonishment for you," he asserts, "then that may possibly be because you are in the habit of treating everything that is beyond your comprehension as a miracle." Diderot would have one explain order without resorting to some notion of preliminary design. It might just as well be simply the consequence of the chance union of elements in an infinity of combinations. The viable ones have persisted while others have necessarily disappeared.[59]

While serving three months in the prison of Vincennes for writing the *Lettre sur les aveugles*, Diderot annotated the first three volumes of Georges-Louis Buffon's *Histoire naturelle*. In an effort to distinguish between brute and living matter, Buffon believed that there are specifically organic molecules scattered throughout nature

[57] Diderot has been the subject of many biographical and philosophical studies. One of the most recent and most satisfying is by Arthur M. Wilson, *Diderot*, New York, Oxford University Press, 1962. A thorough analsyis of the development of Diderot's thought in relation to the life sciences and of physiology is provided by Jacques Roger, 'Diderot et l'Encyclopédie', op. cit., note 7 above, pp. 585–682. A very brief survey is provided by Hall, op. cit., note 41 above, vol. 2, pp. 56–65.

[58] Denis Diderot, 'Pensées philosophiques', in *Oeuvres complètes*, 15 vol., Paris, Le Club Français du Livre, 1969, vol. 1. Roger discusses the work in op. cit., note 7 above, pp. 585–591.

[59] Diderot, 'Lettre sur les aveugles', in *Oeuvres complètes*, op. cit., note 58 above, vol. 2, p. 196. Roger discusses the book in op. cit., note 7 above, pp. 591–599.

which combine to form composite bodies. Diderot's article 'Animal' in the first volume of the *Encyclopédie* expounded upon just that theme. Buffon's work also influenced Diderot's *Pensées sur l'interpretation de la nature* of 1753, in which he speculated about the nature and origin of living species. Echoing Buffon, he wrote that "matter in general is divided into dead matter and into living matter." But that raised other questions. "How can it be", Diderot asks "that matter is not one, either all living or all dead? Is living matter always living? And is dead matter always and really dead? Does living matter ever die? Does dead matter ever begin to live? Is there any assignable difference between dead and living matter other than organization?"[60] Eventually these speculations were to terminate in a notion of unity which rejected as irrelevant the distinction between thought and matter, between life and non-life.

Not satisfied with assigning living phenomena to organization, Diderot wrote in 1759 that it is absurd to say that "particle *a* placed to the left of particle *b* has no consciousness of its existence, does not sense, is inert and dead", while if *b* is to the left of *a*, "the whole lives, knows, senses".[61] By 1765, he had arrived at the breathtaking notion that sensibility is not confined to a living organism. Rather, it is a universal property of matter; an inert property in brute bodies, like movement in heavy bodies stopped by an obstacle; a property activated in the same bodies by their assimilation with a living animal substance.[62] This theme was developed in the *Rêve de d'Alembert* of 1769, a work in which Diderot made Bordeu his mouthpiece. We have already seen that Diderot had contact with various members of the Montpellier school, from whom he commissioned many articles over the years for the *Encyclopédie*.[63] His use of Bordeu, therefore, is a clear assertion that his own notions about sensibility were rooted in the speculations of the Montpellier school.

D'Alembert's dream is preceded by a conversation, *L'entretien entre d'Alembert et Diderot*, in which the two men discuss the question of how apparently brute matter is transformed into that which is living and active. D'Alembert is not convinced by Diderot's theory of the universal sensibility of matter. The work opens with d'Alembert, an exponent of the traditional view of matter, admitting that if, like the dualists, one assigns an external motive force to a living body, one must admit to the existence of a spiritual being that possesses contradictory properties. "I confess to a Being who exists somewhere and yet corresponds to no point in space, a Being who, lacking extension, yet occupies space, who is essentially different from matter and yet is one with matter, who follows its motion and moves it, without himself being in motion, who acts on matter and yet is subject to all its vicissitudes." But to reject the contradictions is to move to Diderot's position, which is also fraught with problems, "for if this faculty of sensation, which you propose as a substitute, is a general and essential

[60] Diderot,'Pensées sur l'interpretation de la nature', in *Oeuvres complètes*, op. cit., note 58 above, vol. 2, p. 770. Roger, op. cit., note 7 above, pp. 599–614. The work is also discussed in Wilson, op. cit., note 57 above, pp. 187–198.

[61] Diderot, 'Lettre de 15 octobre 1759 à Sophie Volland', in *Oeuvres complètes*, op. cit., note 58 above, vol. 3, pp. 815–821. Quoted by Roger, op. cit., note 7 above, p. 617.

[62] Diderot, 'Lettre de 10 octobre 1765 à Monsieur Duclos', in *Oeuvres complètes*, op. cit., note 58 above, vol. 5, pp. 949–952. Quoted by Roger, op. cit., note 7 above, p. 617.

[63] Roger discusses the influence of the Montpellier physicians in the development of Diderot's philosophy of life in ibid., pp. 631–641.

quality of matter, then a stone must be sensitive."[64]

If one acknowledges the existence of sensibility in that stone, then the process of nourishment is understood as one whereby an organism removes obstacles which mask the sensibility. It follows, therefore, that the animists, mechanists, and many vitalists had all committed the same basic error of assuming that life is imposed on to matter such that its properties supersede and dominate the inorganic ones. The truth, Diderot believed, is that life is not imposed but released. When you consume food, he wrote, "you assimilate it, you turn it into flesh, you make it animal, you give it the faculty of sensation."[65]

Agitated by their conversation, d'Alembert later sleeps uneasily. While asleep, he rambles on in a feverish vision which hovers around questions about the nature of living matter, consciousness, and sensation. Called to attend to the apparently delirious man, Bordeu establishes that d'Alembert's pulse and respiration are normal, then settles down to assist Mlle de Lespinasse, d'Alembert's mistress, to extract the implications that follow from the dreamer's visions. Their part in this drama is said to have angered d'Alembert and Mlle de Lespinasse, but Bordeu's reaction is not recorded. In any case, it is not overly difficult to imagine Bordeu expressing many of the ideas Diderot presumed to attribute to him in this work.

Diderot has Bordeu explaining growth and the attendant acquisition of sensibility as follows:

> You begin as an imperceptible speck, formed from still smaller molecules scattered through the blood and lymph of your father and mother; that speck becomes a loose thread, then a bundle of threads. . . . Each of the fibres in the bundle of threads was transformed solely by nutrition and according to its confirmation, into a particular organ. . . . The bundle is a purely sensitive system. . . . This pure and simple sensitivity, this sense of touch, is differentiated through the organs that arise from each separate fibre; one fibre forming an ear, gives rise to a kind of touch that we call a noise or a sound.[66]

Consciousness is merely an organizational refinement of sensitivity confined to the brain, which is the common centre of all the sensations, as sight belongs to the eye and hearing to the ear. With Bordeu's approval, Mlle de Lespinasse draws the appropriate materialist conclusion which is that differences between human beings and animals are finally only organizational. "Where the origin or trunk is too vigorous in relation to the branches, you have poets, artists, imaginative people, cowards, fanatics, madmen. When it is too weak, you get so-called brutes and savage beasts. Where the whole system is slack and soft, without energy, you get imbeciles; where the whole system is energetic, harmonious, well-disciplined, you get sound thinkers, philosophers, sages."[67]

The visionary quality of Diderot's views is exciting if one is pleased with the notion of a universe pulsating with ubiquitous vitality. Matter, it follows, is a panorama of apparently inert but sensitive points conglomerating to form objects some of which are living, conscious beings. Life is continuously emerging from its potential state so that soil, stone, plants, and animals form an intricate and interconnected mass of matter changing its form in such a way that neither birth nor death have any ultimate

[64] Diderot, 'Le rêve de d'Alembert', in *Oeuvres complètes*, op. cit., note 58 above, vol. 8, p. 55.

[65] Ibid., p. 58.

[66] Ibid., pp. 104–105.

[67] Ibid., pp. 134–135.

meaning. As we have always known, everything returns sooner or later to the "great inert sediment" from which it emerged. The *Rêve* is a masterpiece of materialistic vitalism.If one adopts Diderot's thesis, then it follows that the physician is not limited to studying life exclusively with the tools of physics and chemistry. But neither is he sent questing after the essence of an ephemeral and elusive soul or vital principle.

Diderot had moved an immense distance from the iatromechanists whose notions he dismissed as utter nonsense. Anyone who omits sensitivity, irritability, and spontaneity from the calculation of the motion of the sensitive, animated, organized, living beings, he wrote, does not know what he is doing.One day, he predicted, all matter will be demonstrated to have six essential properties – attraction, length, depth, breadth, impenetrability, and sensitivity. The *Éléments de phisiologie* of 1776 was the last work in which Diderot addressed the questions surrounding life and consciousness. His viewpoint remained essentially that which he assigned to Bordeu in the *Rêve*.

Much like La Mettrie, Diderot recognized the life force in a skinned, headless, but still moving snake, in the quivering fragments of an eel, and in the contractions of a pricked, excised heart.[68] Bordeu had written that "the general life . . . is the sum of all the particular lives."[69] Diderot had been much influenced by Bordeu when he commented that in the living organism, there are three distinct levels of life. These are "the life of the entire animal", the "life of each of the organs", and "the life of the molecule". They are inextricably intertwined in the body which they occupy. "The heart, the lungs, the spleen . . . nearly all the parts of the animal live for some time separated from the whole. Even the head separated from the body sees, looks and sees. It is only the life of the molecule, or its sensibility which does not cease at all. It is one of its essential qualities."[70]

Diderot's work is important because it incorporates the most progressive and innovative ideas of his time. Many works, including the *Rêve* and the *Éléments de phisiologie* in which the notion of universal sensibility was explicitly expounded, were not published until 1875. Nevertheless, the new physiology upon which they were based suffused the *Encyclopédie*. It is important, finally, because it appears that until such explicity materialist notions as those of La Mettrie and Diderot were developed, there was virtually no way to avoid calling upon a spiritual, immaterial principle to play some physiological role. Sensitive matter effectively stripped the soul of any physiological functions.[71]

After soaring along with Diderot's imagination it is anticlimatic to return to the somewhat more pedestrian anatomically-based speculations of Bichat. Although Bichat analysed sensibility and contractility and classified them into five distinct vital forces, he had no concept of their universality in matter. Indeed, his definition of life as "the collection of those forces which resist death" precluded such a possibility. His

[68] Diderot, 'Éléments de phisiologie', in *Oeuvres complètes*, op. cit., note 58 above, vol. 13, pp. 661–662.
[69] Bordeu, op. cit., note 29 above, pp. 829–831.
[70] Diderot, op. cit., note 68 above, pp. 662–666.
[71] Ibid., pp. 67–69. An extensive discussion of the connexion between Diderot's and La Mettrie's notions is found in Vartanian, 'L'homme machine since 1748', in op. cit., note 41 above, pp. 114–136.

five forces included the consciously perceived sensibility and contractility described by Haller and the unperceived ones which many persons assumed must occur at the level of organs or even molecules. Most important of all, he insisted that all the vital properties are fundamentally and absolutely different from physical ones.

IV

MEDICINE AND IDEOLOGY: THE METHODOLOGY AND EPISTEMOLOGY OF THE SENSATIONALISTS

Eighteenth-century European intellectuals generally assumed that they were living in a time of fundamental change. They were convinced that intellectual, material, and social progress was possible, that it was occurring in at least some places, and that it was bound to continue. The future must be more enlightened, more free, more tolerant, and more prosperous than the past. Even when they risked imprisonment, vilification, and exile because they disparaged established institutions and attitudes, they did so believing that their notions would finally prevail. It is difficult for us to recreate their optimism. To a large extent, our own cynicism about the present and unease about the future are due to society's failure to realize the *philosophes'* vision Earlier utopian aspirations had been rooted in a Christian ideal and looked towards man's spiritual and moral nature for their realization. The forward-looking man in the eighteenth century, however, believed that a better future would derive from the study of the natural world and of man's place in it. Provided that the study was conducted correctly, man would come to understand both his inner nature and his environment, and be enabled thereby to live harmoniously in the world. In a profound sense, the eighteenth-century savants believed in the virtue of a scientific, technological, and hence materialistic future. These themes have long preoccupied students of the Enlightenment and many persons have written about them at length. My primary interest here is in the way in which this notion of possible and desirable progress affected certain scientists and especially physicians and writers on physiology.

During the eighteenth century, there was a growing sense of the relative backwardness of the life sciences, especially when their achievements were measured against those of physics and chemistry. In the attempt to promote the advancement of medicine, the state intervened, meanwhile offending the hide-bound Paris Medical Faculty, long-time enemies of innovation who sensed their traditional power slipping from them. A Société Royale de Médecine was established by royal patent in 1776 as an agency of the French state, to be consulted by members of the administration. Its mandate was to address questions surrounding human and animal epidemics, sanitary conditions, endemic and occupational diseases, surgical and anatomical subjects, and the supervision of therapeutic mineral waters. It quickly became a haven for the young progressive spirits of French medicine so that, apart from some 162 associate members, it attracted thousands of corresponding members from throughout the kingdom. It was closed in 1793 as an institution of the Ancien Régime. Nevertheless, its members were to contribute much to the medical world of post-revolutionary France as well, bringing to fruition then many of the ideas which had been born before the great iconoclasm.

Although the Société Royale's perpetual secretary, Félix Vicq d'Azyr (1748–94) did not survive the Revolution, many of his ideas found a home in the newly created Écoles de Santé. Remembered especially as a comparative anatomist, he was also much preoccupied with questions of scientific method and direction. He declared that medicine needed to be incorporated into a larger "science of man", one that concerned itself not only with man's physical being but also with his psychological and social make-up. The discussion of just such an integrated science was taken up widely after the Revolution, as we shall see later in this chapter.[1] The Société Royale's optimistic reforming drive and its members' advocacy of a larger "science of man" persisted, finding a good home in the Société Médicale d'Émulation. Bichat, one of its founding members, wrote the 'Preliminary discourse' to the first volume of the *Mémoires de la Société Médicale d'Émulation*, claiming that "the modest reunion of several young friends of the sciences scares some grave persons who appear to suspect there a secretly hatched conspiracy against their antiquated principles and the ruinous state of their doctrine." In a less than conciliatory vein, he said that he hoped that their detractors would abandon envy for reason but, perhaps more importantly, that medicine would be incorporated into a general "science of man". Meanwhile, he invoked Hippocrates as the eternal model to be emulated.[2]

The original membership list of the Société d'Émulation included many of the most famous physicians of the time, many of them former members of the Société Royale. If asked about the philosophy or methodology with which they approached their work, most, if not all, would have claimed to be sensationalists and Newtonians. By that, they would have meant that they undertook to apply to their own work the methods that the great physicist had used in his. The imitation of physics understandably preoccupied many savants, so that the intellectual and scientific world of the eighteenth century was crowded with would-be Newtonians of every conceivable discipline, including theology. Sensationalism was the psychological system, the epistemological theory that, in the estimation of many, accompanied Newtonian theory and provided the theoretical framework for progressive scientific procedure.

Sensationalist theory was rooted in the conviction that learning and knowledge proceed from the five senses and not from some innate pool of ideas. As with many ideas on biological subjects, Aristotle seems to have achieved priority by stating that all knowledge derives from the senses. French sensationalists, however, including most of the members of the Société d'Émulation, traced their ideas on the subject to the Abbé Étienne de Condillac in the 1750s, who, in turn, owed much to John Locke (1632–1704). Locke discussed certain general precepts basic to scientific investigation, which corresponded to Newtonian method and procedure. In 1801, Destutt de Tracy, one of the foremost exponents of sensationalist philosophy, dubbed the system "ideology", and its French proponents came to be generally known as "ideologues".

[1] Caroline C. Hannaway. 'The Société Royale de Médecine and epidemics in the ancien régime', *Bull. Hist. Med.*, 1972, **46**: 257–273. Charles Coulton Gillispie, *Science and polity in France at the end of the old regime*, Princeton University Press, 1980, pp. 194–226. Caroline C. Hannaway, *Medicine, public welfare and the state in eighteenth-century France: the Société Royal de Médecine, Paris (1776–93)*, PhD thesis, Johns Hopkins University, 1974; Ann Arbor, Mich., University Microfilms, 1976.

[2] Xavier Bichat, 'Discours préliminaire', *Mémoires de la Société Médicale d'Émulation*, 1796, **7**: i–xii. For a discussion of the philosophy of the society and especially of its members' views concerning the nature of scientific medicine, see Sergio Moravia, 'Philosophie et médecine en France à la fin du XVIIIe siècle', *Studies on Voltaire and the eighteenth century*, 1972, **89**: 1089–1151.

By the end of the eighteenth century, "ideology" had become very influential in France. The way in which physicians in particular interpreted ideology and adapted it is our primary concern in this chapter. The way in which Bichat's work was influenced by sensationalist precepts should become clear in succeeding chapters.

To early seventeenth-century philosophers and scientists, it seemed that philosophical knowledge is attainable if one proceeds from some original certainty. They presumed that one can correctly deduce all propositions from an initial axiom, so long as it is sound and one proceeds rigorously. Descartes was the most notable spokesman for that position. By the end of the century, this approach seemed much less credible. Newton, for example, took care to affirm that to do his work, he had proceeded not from axioms but from the data of experience and experiment. He elevated the status of data so acquired to a supreme position in science. He also limited the scientists' domain by contending that speculation about metaphysical and theological questions was outside their competence: that is, Newton's approach to nature was inductive. Whereas Descartes' statement of certainty, his "I think, therefore I am", had been his starting-point, Newton's statement of the law of universal gravitation and his formulations about optics were the end-point of his work.[3]

At the end of the seventeenth century, John Locke set out to create a science of the human mind on the basis of Newton's principles. In the words of Jean d'Alembert in the 'Preliminary discourse' to the *Encyclopédie*, "Locke reduced metaphysics to what may be, in effect, the experimental physics of the soul".[4] To do so, he attempted to trace the natural history of human ideas to their simplest form in the consciousness. Locke contended that, lacking innate ideas, the mind is as "white paper, void of all characters, without any ideas". All man's complex mental operations, his reason and knowledge, are the product of the interaction of his sensations and of an innate capacity for reflection. Our senses convey some distinct perceptions into the mind, giving us our ideas of such things as "yellow, white, heat, cold, hard, bitter, sweet, and all those which we call sensible qualities". The reflective capacity of the mind or the soul is a kind of internal sense by means of which we develop "perception, thinking, doubting, believing, reasoning, knowing, willing and all the different actings of our own minds". Acting together, these supply all the material necessary for thought. "These two, I say, viz, external material things, as the objects of SENSATION and the operations of our own minds within, as the objects of REFLECTION, are to me

[3] There are a number of good studies of Newtonianism in the seventeenth and eighteenth centuries. Charles Coulton Gillispie's lively and elegant essay 'Science and the Enlightenment', *The edge of objectivity*, Princeton University Press, 1959, pp. 151–202, remains the first work which ought to be consulted on the topic. Some of the more recent important works are Robert E. Schofield, *Mechanism and materialism. British natural philosophy in an age of reason*, Princeton University Press, 1970; Arnold Thackray, *Atoms and powers*, Cambridge, Mass., Harvard University Press, 1970; T. M. Brown, 'The mechanical philosophy and the "Animal oeconomy" ', Princeton University PhD thesis, 1970; Georges Gusdorf, *Les principes de la pensée au siècle des lumières*, Paris, Payot, 1971, especially 'L'intelligibilité au XVIIIe siècle', pp. 151–289; and finally, the classic work of Ernst Cassirer, *The philosophy of the enlightenment*, trans. by Fritz C. A. Koelin and James P. Pettegrove, 7th ed., Princeton University Press, 1965, pp. 3–36. Gusdorf in particular has examined the importance of this English philosophy and its effects upon French science.

[4] Jean La Rond d'Alembert, *Preliminary discourse to the Encyclopedia of Diderot*, trans. by Richard Schwab, Indianapolis, Bobbs-Merrill, 1963.

the only originals from whence all our ideas take their beginnings."[5]

The Newtonian and Lockian systems took rather a long time to be assimilated in France, for they collided for a time with a patriotically based allegiance to Cartesianism. Indeed, the Cartesian Fontenelle dominated the powerful Académie des Sciences for years, thus discouraging the diffusion of alternate views. Voltaire first assisted the importation of English ideas into France when he cited Locke's views approvingly in his widely read *Lettre sur les anglais* of 1734. He became a powerful propagandist on Newton's behalf, arguing that Descartes had erred fundamentally when he tried to create a physics without reference to experiment.[6] Voltaire's immensely successful *Éléments de la philosophie de Newton* of 1738 was a particularly decisive blow to a Cartesianism that had become something of an albatross. The Lockianism that accompanied Newtonian theory seemed to provide the very conceptual tools which the French needed at the time, and they grasped them eagerly.

The *Abbé Étienne Bonnot de Condillac* (1714–1800) was the first person to develop Locke's theory. He eliminated the sensation-reflection dualism, contending that reflection, like all other faculties, is subordinate to sensation, which is the body's only source of cognition and intellectual activity. It was a bold step insofar as it eliminated any need for a hypothetical pool of innate ideas or for a divine origin for rational thought. Condillac also propounded in detail a method of scientific investigation known as "analysis", which was supposed to be the inductive method of Newton. In time, scientists and philosophers came to see the analytic procedure as the indispensable element of all labour of the mind. The political, moral, and biological applications of Condillac's work were to preoccupy many French thinkers, scientists, and politicians alike, until at least the post-revolutionary period. Its implications for medicine and physiology were examined and systematized in the writings of Tracy, Cabanis, and Pinel. Their work, in turn, strongly affected that of Bichat. The fundamental philosophy of Condillac, if not necessarily its details, was to be the reference point for all those persons who, like the members of the Société d'Émulation, wished to study the whole man in his psycho-physiological complexity.

The "philosopher of the *philosophes*", as Condillac was known, was an unusual man insofar as the Jesuits commended his work as much as did the materialists. A circumspect man and an ordained priest, he expressed the most potentially radical ideas in such a way that his lifelong loyalty to the Catholic faith never came into question. Étienne Bonnot took the name Condillac from the Château de Condillac which his father, a member of the petty nobility, had purchased in 1720. Condillac studied theology at the Sorbonne. Reserved, cold, and humourless, he knew the sophisticated salons of France, but did not shine in them. He became friends with Diderot and Rousseau in Paris while they were still all literary unknowns.[7] In his *Confessions*, Rousseau related that it was Diderot who found a publisher for Condillac's first book,

[5] John Locke, *An essay concerning human understanding*, New York, Meridian, 1964, bk. II, ch. I, pp. 89–98.

[6] [François-Marie Arouet] Voltaire, 'Sur M. Locke', in *Lettres sur les anglais*, Cambridge University Press, 1931, pp. 45–51.

[7] Isabel F. Knight, *The geometric spirit; the Abbé de Condillac and the French enlightenment*, New Haven, Conn., Yale University Press, 1968, pp. 1–16.

his *Essai sur l'origine des connaissances humaines* of 1746. Meanwhile, he claimed for himself the distinction of being the first person to have appreciated Condillac's remarkable abilities.[8]

Condillac wrote the *Essai sur l'origine*, he claimed, to introduce Locke to the French public. By observing how sensations and impressions combine to produce knowledge and intelligence, he hoped to establish a philosophy of knowledge whose principles and procedures would in themselves constitute an exact science.[9] Contending that all knowledge derives from sensory perception and that all complex perceptions are compounds of simple ideas, he imagined a newborn child to be but a passive recipient of a sudden multitude of confused sensations. In time, that which is first experienced as merely a bombardment of light and colour, discomfort and pleasure, motion and sound is developed into perception, consciousness, attention, memory, imagination, and contemplation.[10]

In 1754, he supplemented that work with the *Traité des sensations*, in which he specifically denied the existence of the reflective capacity which Locke had conjectured co-existed with sensation. He asserted that the psychic life and all its faculties and passions are developed from simple sensation alone.

To illustrate his thesis, Condillac described a hypothetical statue in which he could juggle the number and combination of the five senses. He began his examination of the process of learning by endowing it with smell, which he described as the least intellectual of all the senses. Since the statue, like man, is deprived of innate ideas, he wrote, the statue can know nothing except what it learns from odours. He undertook to show how, with smell, it would build up a knowledge of itself and of the world external to it, how it would develop its passions and complex abstract notions, and even how it would develop a notion of morality and of a deity.[11] This use of a fictional statue as a kind of thought experiment was, at the time, a fairly common device. Buffon, for example, wishing to give an account of the development of human thought, extrapolated Locke's notions and speculated about a kind of "epistemological Adam". Confused initially by the bombardment of sensory impulses, the hypothetical first man learned eventually to distinguish ideas and to think. And in 1751, just three years before the appearance of the *Traité des sensations*, Diderot composed a *Lettre sur les sourds et muets à l'usage de ceux qui entendent et qui parlent*, in which he suggested the possibility of reconstructing the formation of

[8] The petulant and testy Rousseau took credit for Condillac's first success: "The booksellers of Paris are always arrogant and hard towards a new author, and metaphysics, which was not much in fashion at the time, did not offer a very attractive subject. I spoke of Condillac and his work to Diderot, and introduced them to each other. They were made to like each other and did so. Diderot induced Durant the bookseller to accept the Abbé's manuscript, and this great metaphysician received for his first book, and that almost as a favour, one hundred crowns and even that he would perhaps not have received but for me." *Confessions of Jean-Jacques Rousseau*, ed. by Lester G. Crocker, New York, Washington Square Press, 1956, bk. 7, pp. 171–172.

[9] Étienne Bonnot de Condillac, 'Essai sur l'origine des connaissances humaines', in *Oeuvres complètes*, 23 vols., Paris, Ch. Houel, 1798, vol. 1, pp. 1–16. See also the elegant work by Gerd Buchdahl, *The image of Newton and Locke on the Age of Reason*, London, Sheed & Ward, 1961.

[10] Knight, op. cit., note 7 above, pp. 17–51. In this second chapter she examines the goals and the assumptions behind the *Essai*.

[11] See Condillac's 'Extrait raisonné', in 'Traité des sensations', *Oeuvres complètes*, op. cit., note 9 above, vol. 3, pp. 3–46.

human thought. In his particular thought experiment, Diderot imagined five persons, each in possession of a difference sense, relating their particular experiences and perspectives.[12]

How does Condillac's statue, possessing only the sense of smell, come to learn something of itself and of its world? A first odour strikes it and engages its *attention*. It experiences pleasure or discomfort depending upon the nature of that odour. When the statue encounters a second odour, its sensory capacity comes to be shared between *memory* of the first odour and attention to the present one. With a succession of odours, the statue learns to *compare*. If it attends to the odour of, say, a rose and a carnation, it perceives that there is a difference between them, from which comparison springs *judgement*. Eventually, to compare and to judge become a habit. Our statue was not surprised at the first sensation it encountered. It became capable of surprise only when it passed suddenly from an accustomed state to a new one. Surprise augments the activity of the operations of the soul, a word which Condillac used in the traditional sense to denote mind or consciousness.

As the soul compares past and present sensations, it develops a *need* which is felt as a privation of a desirable sensation. Our statue is capable of developing the faculty of *imagination*, which is really a more vivid form of memory. By now, our statue has developed a considerable number of faculties. Finally is born *desire*, which Condillac considered to be a major moving force behind human mental development. He described it as the faculties directing themselves to something for which one senses a need. To experience desire, a statue must be able to judge the difference between an existing state and a remembered one so as to be able to determine which is more acceptable. Thus are born the passions such as joy, suffering, need, desire, love, and hate. So far, however, the love is restricted to itself, for the statue is not yet aware of an external world.[13]

At this stage, Condillac continued, the statue should be capable of forming abstract ideas. Since the odour of each flower is particular, our statue will come to discern number from a succession of odours. It will learn of coming and going, of beginning and end, of past, present, and future.[14] He wrote that, "if we consider that to recall, compare, judge, discern, imagine, to be astonished, to have abstract ideas, and to have number and duration, to know general truths and particular ones are nothing but different ways of being attentive; that to have passions, to love, to hate, to hope, to fear and to want, are but different ways of desiring: and that finally to be attentive and to desire are really but to sense; we conclude that sensation encompasses all the faculties of the soul."[15]

A statue possessing only the sense of hearing would develop similar faculties to the statue which could but smell. A statue in possession of two senses would acquire the same faculties extended further. The statue learns to perceive form with the eye only because of that organ's unique structure. It would do so with the nose also if the

[12] Gusdorf, op. cit., note 3 above, pp. 232–249. Diderot's relationship with Condillac and with La Mettrie is discussed by J. H. Brumfitt, *The French enlightenment*, London, Macmillan, 1972, pp. 99–132.

[13] Condillac, op. cit., note 11 above, pp. 56–95.

[14] Ibid., pp. 96–120.

[15] Ibid., pp. 121–122.

particles that produce the sense of smell travelled in straight lines, and if the nose were so constructed as to accept their input in parallel lines. Neither sight, nor smell, nor taste, nor hearing, however, will allow our statue to perceive that there exists a world external to itself. A sense of touch is essential to permit a living being to experience the existence of its own material body and therefore to learn that there are bodies besides itself in the world.[16]

Condillac believed that even the religious experience has a sensory foundation. The fact that the human race universally possesses the idea of God has long been the major argument for the existence of innate ideas. But Condillac believed that the religious sense comes once a human being learns that he is not alone in the world, and once he begins to sense his dependence. A child believes for a while that what is pleasing exists merely to please him and that which is disagreeable exists only to cause him pain. A prayer is born of the desire for whatever is favourable to his senses. Thus it is that primitive man, nearer to a childlike state than civilized man, addresses himself to the sun to urge it to shine and to the trees to urge them to bear fruit; he treats agents of pain and discomfort as disagreeable enemies to be appeased. And so the universe becomes populated with visible and invisible beings. The sophisticated mind proceeds from these first divinities to a more complex notion of first cause and thus to montheism.[17]

In the *Traité des sensations*, Condillac went on to discuss the analytic method, which derived from a sensationalist perception of the mind's functions. It has to do with comprehending complexity by dissociating the component elements of a thing so as to study them separately. By sorting out its parts, one imposes an order on to apparent chaos.[18] In his *Logique* of 1780, Condillac illustrated analysis with a country view. If we examine it merely as an attractive composite, we learn nothing substantial about it. To do that we must decompose it so as to examine its parts:

> If [nature] gave us the faculty of seeing a multitude of things at once, it also gave us the faculty of looking at but one thing, that is to say, of directing our eyes to one only; and it is this faculty which is the result of our organization, which gives us all the knowledge which we can acquire by the use of sight . . . . Thus one begins with the principal objects: one observes them in succession and compares them to judge their connexions. When by this means, one has their respective situation, one observes in succession all those things in between them, or compares each of them with the nearest principal object and so determines their position.[19]

Thus it is that one comes to understand a countryside, universal gravitation, or the functioning of the human body. A resynthesis completes the analytic process. "We decompose only in order to recompose; and when knowledge is acquired, things,

[16] Ibid., pp. 124–272.

[17] Ibid., pp. 388–391. The question of how man acquires a knowledge of God is dealt with at somewhat greater length in Condillac's 'Traité des animaux', in *Oeuvres complètes*, op. cit., note 9 above, vol. 3, pp. 565–586.

[18] Condillac, op. cit., note 9 above, pp. 157–172. Analysis was described as the method appropriate to scientific inquiry by Isaac Newton, who wrote in the *Opticks* as follows: "We may proceed from compounds to ingrediants [*sic*] and from motion to the forces producing them, and in general from effects and their causes and from particular causes to more general ones, till the argument end in the most general. This is the method of analysis; and the synthesis consists in assuming the causes discovered and established as principles, and by then explaining the phenomena proceeding from them and proving the observations." Quoted by Buchdahl, op. cit., note 9 above, pp. 69–74.

[19] Condillac, 'Logique', in *Oeuvres complètes*, op. cit., note 9 above, vol. 22, pp. 16–20.

rather than being successive, have the same order in the mind that they have outside. Our knowledge consists in this order."[20] Condillac assumed that he had employed just such an analytic process, albeit imaginary, to demonstrate that all knowledge is sensory. But as Knight has pointed out, he is open to the criticism that he violated his own methodological principles because, instead of commencing a study with verifiable facts, he used a fictitious statue-man.[21]

In his *Logique*, Condillac addressed the subject of language, which he described as coinciding with analysis itself. Once again, he was taking up a theme popular in the Enlightenment. Many persons speculated about the nature of original language and its connexion with human consciousness. In 1752, for example, Maupertuis addressed a *Lettre sur le progrès des sciences* to Frederick II, proposing some experiments on, for example, two or three abandoned children to isolate them complete from the external world so as to observe their language development. A good example of the fruit born of Condillac's treatment of the subject is the achievement of Antoine Lavoisier, who assiduously studied his work and subsequently produced a linguistic as well as a methodological reform in chemistry. In the eighteenth century, of course, chemistry was rudimentary, lending itself to analysis because it was concerned with complex matter constructed from more elementary matter.[22]

Condillac acquired considerable prestige and influence in his own lifetime. His philosophical achievement was honoured not only by scientists and *philosophes*, but by various sections of government as well. From 1758 to 1767, Condillac was instructor to the Duke of Parma, nephew of Louis XV. Later, he turned down an opportunity to teach the three sons of the dauphin. A *Cours d'études* which he composed largely for the instruction of the Duke of Parma contained sections on grammar, the art of writing, the art of reasoning, the art of thinking, and a general history of man with a philosophy of history. In 1769, he received a place in the exclusive Académie Française. In 1777, the Polish government invited him to write an elementary logic for use in their public schools. This was, in fact, the source of his *Logique*, an outline of his ideological methods described for purposes of public instruction. It was treated as a model for education in French schools until shortly after the Revolution.

One of the first persons to become involved with Condillac's ideas and to publicize them was *Jean le Rond d'Alembert* (1717–83), an accomplished mathematician and physicist. He was one of the first Frenchmen to delve into the implications of Newtonian science and philosophy, and his *Treatise on dynamics* of 1744 is generally taken to be a landmark in Newtonian mechanics. He accepted, more or less as given, the sensationalist point of view which, for many, accompanied Newtonianism. It was d'Alembert who composed the "manifesto of the French Enlightenment" and its "discourse on method", the *Preliminary discourse* to the first volume of the *Encyclopédie*.[23]

[20] Ibid., pp. 12–22.

[21] Knight analyses Condillac's 'Traité des sensations' in op. cit., note 7 above, pp. 52–78.

[22] Gusdorf, op. cit., note 3 above, pp. 232–249. Condillac's work and the question of language and scientific nomenclature are discussed by Moravia, op. cit., note 2 above.

[23] Schwab, 'Translator's introduction', *Preliminary discourse*, op. cit., note 4 above, pp. ix–xxxi.

Close friends since the 1740s, d'Alembert and Condillac must have influenced one another's notions of scientific method and epistemology. Indeed, Condillac's influence on the *Preliminary discourse* is unmistakable. At the beginning of the work, d'Alembert asserted that "All our direct knowledge can be reduced to what we receive through our senses; whence it follows that we owe all our ideas to our sensations." He described how our sensations teach us, first, the fact of our existence and then, that of external objects in which he included our bodies. Differing somewhat from Condillac, however, d'Alembert described three faculties of the mind that react with the sensations – memory, which is the recalling of sensations; reason, which involves comparing; and imagination, which is the creation of new ideas.

D'Alembert also described the process of analysis, which he assumed produced seventeenth- and eighteenth-century mathematics and physics.[24] He discussed the process again in the article 'Analytique', which he composed for the first volume of the *Encyclopédie*. He wrote, in imitation of Newton, that the essence of scientific method is the path by means of which "one may proceed from composite substances to their elements, from motions to the forces which produce them, and in general, from effects to their causes, and from particular causes to more general ones, up to the point where one arrives to that which is the greatest of all."[25] Through the *Encyclopédie*, Condillac's basic philosophy reached what was probably the widest audience possible at the time.

One of the first physicians to pay special attention to Condillac's texts was *Felix Vicq d'Azyr* (1748–94), who came to Paris from Normandy to study medicine in 1765. One of those enviable persons who is able to embrace a wide range of subjects and to enrich one by borrowing from another, he contributed substantially to contemporary discussions of anatomy, physiology, medical theory, and hygiene. Vicq d'Azyr tried consciously and consistently to apply Condillac's notions to his work and teaching. With analytical principles in mind, he developed the discipline of comparative anatomy by deliberately studying organs and functions in isolation, and then linking together as units.[26] As permanent secretary of the Société Royale de la Médecine, he helped to make the most innovative French physicians conscious of sensationalist and analytic principles and of their potential application to a reform of medical teaching and practice.

On behalf of the Société Royale, Vicq d'Azyr presented certain reflections on medical teaching to the French National Assembly in 1790. The document was a powerful indictment of a profession which he described as populated with ignorant physicians and charlatans. In the entire country, he charged, there was no place where the principles of the healing art were adequately taught. Entry into medical schools was too easy, and once inside, the student received insufficient instruction and was

[24] d'Alembert, op. cit., note 4 above, pp. 6–11.

[25] d'Alembert, 'Analytique', in Denis Diderot, *Encyclopédie*, 35 vols., Paris, 1751–80, vol. 1, pp. 403–404.

[26] Vicq d'Azyr's work has not, thus far, been extensively studied. The most useful biography I have found is Jacques Moreau's 'Discours sur la vie et les ouvrages de Vicq d'Azyr', *Oeuvres de Vicq d'Azyr*, 6 vols., Paris, L. Duprat-Duverter, 1805, vol. 1, pp. 1–88. Vicq d'Azyr's general role in eighteenth-century French medicine is discussed by Moravia, op. cit., note 2 above. Moravia deals with Vicq d'Azyr's ideological convictions, his link to the Société Médicale d'Émulation, and his wish to see the medical profession develop autonomy.

subjected to inadequate examination. Medical theory in general, unable to free itself from its dependence upon other sciences, was unable to develop its own instruments, techniques, and laws.[27] Vicq d'Azyr argued that medicine ought to define itself much more broadly, going beyond the mere examination of bodily parts and functions. Its preoccupations should also include psychology and even human social organization and its institutions.[28] Vicq d'Azyr placed man at the summit of a chain of living beings and invited the philosopher, savant, artist, and literary man to join the physician and naturalist in studying him.[29] Some of his notions were presented to the Revolutionary convention by Antoine Fourcroy shortly after Vicq d'Azyr's death. Thus he contributed towards the reformed education system which produced the Écoles de Santé in 1795.

Vicq d'Azyr's outspoken demand for autonomy and reform was not a new theme. It was implicit in the Montpellier physicians' reaction to iatromechanism and to the subordinate status of life sciences that it implied. Montpellier had considerably influenced Parisian intellectual life through such persons as Bordeu, Barthez, and Fouquet.[30] By the time Vicq d'Azyr was writing, physiology had developed a language appropriate to its concerns. Addressing itself to sensibility and contractility, it no longer had to adopt an imagery of mechanics and of forces belonging to physics. It had a basis upon which to develop a mature and separate discipline. The vision of Vicq d'Azyr and the sensationalists incorporated more than physiological theory, however. It also had to do with medical practice, hygiene, and psychology, such that it was a broadly defined healing art secure in its status as an autonomous science and confident in its premisses. The scientific method of analysis and the sensationalist psychology upon which it was based had much to do, therefore, with the post-revolutionary changes in medical teaching and practice.

One salon in particular, that of Madame Helvétius at Auteuil near Paris, became an important centre for the elaboration of Condillac's principles. Before the Revolution, Madame Helvétius entertained such famous men as d'Holbach, Diderot, d'Alembert, Volney, Lavoisier, and the Americans, Jefferson and Franklin. Cabanis and Tracy, who belonged to a kind of third generation of *philosophes*, had certain of their ideas shaped by that company. For a time, the ideologues in general were considered to belong to an exclusive company of intellectuals who were able to contribute substantially to government and reform. They had tended to welcome the French Revolution initially, being generally interested in questions of politics, education, and so on. They generally agreed upon the necessity to create rational and free institutions to

[27] Felix Vicq d'Azyr, 'Reflexions sur les abus dans l'enseignement et l'exercise de la médecine', *Oeuvres*, op. cit., note 26 above, vol. 5, pp. 57–67. For a discussion of the general situation of the medical profession and the context in which Vicq d'Azyr's proposals were made, see Paul Delaunay, *Le monde médical Parisien au dix-huitième siècle*, 2nd ed., Paris, Jules Rousset, 1906, especially pp. 21–27. Michel Foucault, *The birth of the clinic*, trans. by A. M. Sheridan, London, Tavistock, 1973, pp. 64–87, discusses the theme of a new medicine based upon the clinic and of Vicq d'Azyr's notion of hospital teaching as the solution to many of its problems.

[28] Vicq d'Azyr, 'Idée générale de là médecine et de ses différentes parties', *Oeuvres*, op. cit., note 26 above, vol. 5, pp. 44–57.

[29] Vicq d'Azyr, 'Exposition des caractères qui distinguent les corps vivants', ibid., vol. 4, pp. 229–312.

[30] For a later statement of that position it is interesting to read Alexis Alquie, *Précis de la doctrine médicale de l'École de Montpellier*, 4th ed., Montpellier, Frères Ricard, 1846, especially pp. 27–102.

replace those of the rigid and anachronistic pre-revolutionary regime. In the 1790s, much was to change, and Auteuil became a refuge for certain liberal intellectuals who were repelled by the Terror and by the excesses of radical politics in general. And again, after 1801, it sheltered ideologues rejected by Napoleon and a newly reconstituted French establishment.

The systems of Locke and Condillac easily lent themselves to the liberal political philosophy with which the ideologues became identified. According to Condillac, as we have seen, much of an animal's development derives from its perception of sensations as being either pleasurable or uncomfortable, and from its striving after the one and the avoidance of the other. The pleasure-pain principle was extrapolated by some into political notions having to do with liberty and the basis of justice. If one is able to exercise one's will in conformity with a desire, one presumably possesses at least some measure of freedom. The pursuit of pleasure and the avoidance of pain can be and often were, in the eighteenth century, made the sole guides of moral judgement. Such an approach, presumably unencumbered by tradition, superstition, and habit, was judged to be more "natural" and hence closer to some kind of pristine human state. The intention of the ideologues who derived their political assumptions from psychological ones was to permit Frenchmen to create a moral and social order based upon utilitarian and materialistic notions of man and society. Education, they said, should be directed to that end and to the destruction of superstition and prejudice by the teaching of science. After the fall of Robespierre, many ideologues functioned as enthusiastic legislators and reformers of French institutions until 1801. That year, the motto "Dieu et l'Empereur" signalled the rise of Napoleon Bonaparte to undisputed power in France. As the motto suggested, a purely secular morality and the teachings of social utility were no longer to be countenanced.[31] Napoleon turned against his former friends and supporters and even had them suspended from the Institute, thereby forcing them to form their centre of intellectual opposition at Auteuil.

One of the habitués of Auteuil as well as a refugee there after 1801 was *Antoine-Louis-Claude, Comte Destutt de Tracy* (1754–1836), a philosopher and politician. Among his contemporaries, he came to be acknowledged as one of the most important of all the exponents of sensationalist philosophy. Under threat of the guillotine, Destutt de Tracy studied the sciences in Carmes prison, where he was thrown on charges of incisivism and aristocracy. His fellow-inmate Lavoisier introduced him to the works of Locke and Condillac. Named a senator in 1799 for his support of Napoleon's *coup d'état*, he went into a kind of exile in Auteuil in 1801, when he had time to put thoughts to paper.

Destutt de Tracy dreamed of creating an entire philosophical system that would include politics, economics, morals, ethics, language, and the sciences. As it was initially conceived, it was the first major effort to produce a theory of human psychology and epistemology tied to a study of the economic, social, and political forms of society. He labelled the all-embracing new science "ideology". But he abandoned

[31] Charles Hunter Van Duzer, *Contributions of the ideologues to French revolutionary thought*, Baltimore, Md., Johns Hopkins University Press, 1935, p. 14. For a discussion of the ideologues' education theories, see L. Pearce Williams, 'Science, education and the French Revolution', *Isis*, 1953, **44**: 311–330.

the *Éléments d'idéologie* when his friend Cabanis died, after completing sections on *Grammaire générale, Logique*, and *Traité de volonté*.

Destutt de Tracy claimed to be addressing young people "to make you see in detail what happens when you think, speak and reason." With the sort of purportedly rational zeal that sometimes afflicts persons preoccupied with science, he began by warning his readers to beware of those philosophers, "amiable enchanters but very dangerous seducers," who, like poets, reason according to their imagination and not after the facts.[32] In asking the broad question, "What is thought?", Destutt de Tracy arrived at a system that differed somewhat from those of Locke and Condillac. Although he agreed that the first faculty of the body is sensation, he did not believe it to be the only one. Our thoughts are composed of the interaction of four elementary faculties. To sensation (I feel a burn) he added memory (it occurred yesterday); judgement (a body burned me); and will (I desired to remove myself from that body). Ingenious though it was, Condillac's imaginary statue was deemed unsuitable for explaining how a human being learns, simply because it lacked internal organs, receiving all its impressions from the exterior of the body. A living body, on the other hand, also experiences sensations of colic, nausea, hunger, pleasure, and pain.[33]

Like Condillac, Destutt de Tracy set out to examine rationally how one learns, composing complex ideas from simple ones.[34] His approach reminds one often of that of his mentor. A significant addition to the development of sensationalism at the hands of Destutt de Tracy is an extended discussion of habit, which he considered to be basic to all learning. It is by means of habit that one learns to dance, to play a musical instrument, to read, and so on. It affects even memory and judgement. The more often something is experienced, the less one must consciously decompose its elements, and the less acutely one feels the sensations it produces.[35] Again like Condillac, Destutt de Tracy considered language to be the primary and even the ultimate intellectual tool. It is the mechanism for all operations of thought, an analytical tool guiding human intelligence and permitting the expression of ideas.[36]

The ideologues' preoccupation with language can be illustrated from the work of Charles-Louis Dumas of Montpellier, who, in 1797, published his *Système méthodique de nomenclature et classification des corps humains*. He hoped to suggest how parts of the body might be named, preferably on the basis of their points of muscle attachment, so as to convey maximum information. Complicated and unwieldly, the result remained deservedly obscure. His more coherent and influential *Principes de physiologie* of 1801 opened with a statement on sensation and analysis, the 'General principles for the good philosophical method in the study of the sciences'. Although it reveals Condillac's almost inevitable influence, it also shows us that Dumas reverted to Locke's notions that ideas are the products of external sensations

[32] Antoine Louis Claude Destutt de Tracy, 'Idéologie proprement dite', *Éléments d'idéologie*, 3rd ed., Paris, Courcier, 1817, p. 19.

[33] Ibid., pp. 22–39.

[34] Ibid., pp. 78–92.

[35] Ibid., pp. 274–294.

[36] Ibid., pp. 309–323. For more material on Tracy, see François Picavet, *Les idéologues*, Paris, Felix Alcan, 1891; Van Duzer, op. cit., note 31 above; and especially Emmett Kennedy, *Destutt de Tracy and the origins of ideology, Mem. Amer. Phil. Soc.*, 1978, no. 129.

reacting with internal reflection.[37]

The *philosophes* largely scorned the European Middle Ages and tolerated the Renaissance only slightly better. Peter Gay has shown how deep was their veneration, however, for the pre-Christian civilizations of Greece and Rome, which most of them judged to have been genuinely progressive because they were free of religious superstition.[38] Among those ideologues who were physicians, there was a particular flowering of interest in Hippocratic medicine. Because he lived a very long time ago and in a somewhat different scientific ethos, Hippocrates could be all things to all men. For centuries, he had been every physician's symbol of the ideal physician, as much as the interpretation of that ideal altered. He was important for the French vitalists, the members of the Société d'Émulation, and the ideologues in the eighteenth century because they took his work to be a model of good method. They frequently affirmed that Hippocrates was a great physician because he worked basically in the way his modern successors were advocating. He observed, he analysed, and he tried to understand an illness not merely in terms of its individual symptoms but in terms of the whole man and of the circumstances in which he found himself. To account for the state of health and for human temperament, for example, Hippocrates took serious account of such variables as age, sex, temperament, regimen, profession, and climate.[39]

Jean-Georges Cabanis (1757–1808), the ideologue physician *par excellence*, frequently referred to Hippocrates in his medical writings. At the age of ten, Cabanis attended the Collège de Brives, where he was taught by a man who tried consciously to apply the analytic method to the study of languages and grammar. After a disappointing flirtation with the Greek literary classics, Cabanis turned to medicine, qualifying as a physician at Rheims in 1784. As early as 1778, he began to attend the salon at Auteuil, where he became friends with Destutt de Tracy. Between 1785 and 1789, he lived there as a kind of adopted son of Madame Helvétius, and he inherited her home in 1800.[40]

He demonstrates well the relationship between sensationalism, medicine, and the science of man. His first major work, *Degré de la certitude de la médecine*, 1788, examined the question of what is good scientific method in medicine. To acquire knowledge about life, death, and illness, he argued, it is sufficient to observe and to examine sensible phenomena and their interconnexion. While it is the observer's duty to study carefully events and their connexions, he must remember that the principles of life, the ultimate nature and cause of disease, and the nature of healing substances are all secrets beyond the capacity of researchers. One must remain within the limits of sensory phenomena. That was not to deny the possibility of medicine's becoming a rigorous science. It was, however, to deny the validity of systems and of a priori principles.

[37] Charles Louis Dumas, *Principes de physiologie*, 4 vols., Paris, Deterville, 1800–04, vol. 1, pp. 8–9.

[38] This is the theme of Peter Gay's *The Enlightenment, an interpretation*, New York, Vintage, 1968.

[39] Moravia, op. cit., note 2 above, pp. 1098–1105.

[40] Biographical details are available in Claude Lehec's 'Biographie', *Oeuvres philosophiques de P. J. G. Cabanis*, ed. by Claude Lehec and Jean Cazeneuve, 2 vols., Paris, Presses Universitaires, 1956, vol. 1, pp. v–xxi; and in Picavet, op. cit., note 36 above, pp. 176–224.

Cabanis' thoughts on the subject of analysis or scientific method as it applied to medicine were expounded further in his *Coup d'oeil sur les révolutions et sur la réforme de la médecine*, 1804. The years between its writing and that of the *Certitude de la médecine* were largely taken up with politics. A firm believer in liberal reform, Cabanis joined public life after the execution of Robespierre. He sat on the Hospital Commission after the re-establishment of medical education; he was appointed Professor of Hygiene and Clinical Medicine in Paris; and he became a member of the National Institute and was made a representative on the Council of Five Hundred. Like Destutt de Tracy, he was named a senator as a reward for his support of Napoleon's *coup d'état* in 1799. In 1804, however, he was deprived of membership in the National Institute and the Senate, and he retired to Auteuil to write.

In *Révolutions et réforme*, Cabanis discussed his ambiguous feelings toward the profession of medicine. It was currently like a traveller hindered with excess baggage, which represented the great weight of material collected without judgement or discrimination. In spite of medicine's dismal past, however, its potential was enormous, if men would but undertake to be governed by reason.[41] The Greeks, of whom Cabanis wrote in superlatives, were taken to be an appropriate model. Hippocrates, he alleged, had delivered medicine from false systems and had created new methods for it. The analytic method is basically Hippocratic procedure recently rediscovered.[42] Francis Bacon was the first modern man to acknowledge the supreme importance of sensory evidence. Thereafter, many persons including Hobbes, Locke, Bonnet, and especially Condillac had progressively refined the process of philosophical analysis, making it surer and simpler.[43]

Cabanis' description of analysis was much like that offered by Dumas in his *Principes de physiologie*. Both men were influenced by the Baconian inductive method, which they fused on to sensationalism. Cabanis distinguished four basic types of analysis. An *analysis of description* examines a body's size, shape, form, and the relation of the parts. An *analysis of decomposition or recomposition* is involved, for example, in trying to discover the workings of a watch by taking it apart and reassembling it. This was presumably the method that Lavoisier and Bichat employed to do their chemical work and tissue work respectively. A *historic analysis* is an examination of how phenomena succeed one another. It is suitable for understanding plant growth, muscular action, or the progress of an illness. It was, in fact, Hippocrates' method for observing disease. Finally, the *analysis of deduction* considers the ideas we receive from objects rather more than the mere objects themselves. With it, one can compare ideas, determine their interconnexions and thereby deduce a new series of truths. By a series of transformations, one can arrive at some primary ideas that cannot be verified by experiment. For example, the deduction that the sun will rise tomorrow belongs in this category. So does the analysis of drugs and the determination of their effectiveness.[44] Such was Cabanis' faith in sound

[41] Pierre Jean Georges Cabanis, *Coup d'oeil sur les révolutions et sur la réforme de la médecine*, Paris, Crapart, Caille et Ravier, 1804, pp. 3–35.

[42] Ibid., pp. 72–97. See also Pierre Jean Georges Cabanis, *Rapports du physique et du moral de l'homme*, 2 vols., Paris, Bechet Jeune, 1824, vol. 1, pp. 11–36.

[43] Cabanis, op. cit., note 41 above, pp. 145–194.

[44] Ibid., pp. 270–313.

method that he believed that twenty-five or thirty years of reorganizing and reclassifying the facts would suffice to verify all observations except perhaps those having to do with epidemics.

Cabanis' principal work was his *Rapports du physique et du moral de l'homme*, which appeared in 1805. It dealt with questions of human consciousness and learning and has led people to describe the author as a precursor or even the founder of psychosomatic medicine.[45] Whether or not one accepts so enthusiastic an evaluation, the work was Cabanis' particular contribution to the goal of a unified science of man. Human nature, he claimed, must be studied as a composite of man's physical and moral make-up. Medical science must, henceforth, take into account man's physical, cultural, and social milieu, as well as the condition of his physical body. Environment invariably affects the condition of a body, the type of illness to which it is subject, and its reactions to medication and other treatment. To be totally effective, the physician must also be a moralist – we might say a psychologist – who seeks to perfect the private life, a legislator who tries to correct a national or social situation, and a healer of the various problems that affect man's wellbeing in general. Cabanis' political involvement, therefore, was entirely consistent with medical practice.

The notions expounded in *Du physique et moral* were grounded in sensationalism. Life, Cabanis wrote, consists in a series of movements or activities executed in response to the impressions received by various organs. Without our senses, we in effect would not exist as living beings, for to live is to feel or to sense.[46] Indeed, Cabanis described birth as not merely a change but a completely new life for the animal. A foetus possesses only the sense of touch, because that alone can be of some small use to it. Only when it leaves the womb, however, does it begin to develop sight, taste, smell, and hearing, and thus to live its full sensory life.

Like Destutt de Tracy, Cabanis found Condillac's statue-man to be too limited to describe learning adequately. Although the effects of sensations in the animal are primary, they are modified by such variables as age, temperament, health, climate, and regimen. Some of these factors are external in origin, while others are purely internal. Habitual ideas and feelings are often altered by factors that have nothing to do with external sensations. Puberty provides us with perhaps the best example of a major series of such changes. Perhaps less dramatically, even the digestive organs, the heart, and the great vessels can have profound effects upon us. In spite of his affirmations of loyalty to Condillac and to Locke, Cabanis found it necessary to reintroduce at least some innate principles into the body. It is necessary, he believed, to postulate an instinctive life that coexists with the sensory life to account for certain unlearned visceral phenomena. This innate life was not, however, capable of accounting for such phenomena as the human religious sense. Indeed, according to Cabanis, it is absolutely unintelligent, only touching the conscious mind at the level of strong emotions and responses. Unlike Condillac's theory, it can account for sexual instincts, a maternal sense, and for such instinctive activity as nest-building among birds, as well

[45] Pierre Astruc, 'Les idées de Cabanis sur la médecine', *Le progrès médicale*, 1956, **84**: 372–374.
[46] Cabanis, op. cit., note 42 above, vol. 1, pp. 36–42.

as for all other activities which have nothing to do with habit or judgement.[47]

The reader will recognize here a variant on Grimaud's and Bichat's separation of an animal's life into external and internal categories. Whether consciously or not, Cabanis was here echoing Bichat's contention, developed some four years earlier in *La vie et la mort*, that birth is the beginning of an external life which derives from conscious sensations. In the shelter of the womb, Bichat said, only internal or organic functions can exist. In that condition, an animal's consciousness and intelligence cannot develop, so that a foetus is little more than a kind of a plant.[48]

Cabanis' notions of sensationalism and psychology had important social implications. As long as one assumes that all human beings receive the same impressions from the external world through sensory organs which behave in a consistent way, one may believe in the possibility of achieving human equality. Its possibility, in fact, seems to be implied in Condillac's viewpoint. According to Cabanis, however, people vary widely not only as a result of their environment, but also because of such factors as age, sex, state of health, and so on. Indeed, they are born neither physiologically, temperamentally, nor intellectually equal, making it impossible for them to learn and develop in similar ways.[49]

This preoccupation with man as a complex amalgam of mind and body shaped by physical, psychological, and environmental factors received considerable attention in the eighteenth century. Hippocrates had probably pointed the way to this type of study in *On airs, waters and places*, an analysis of the effects of weather, climate, the seasons, and the landscape on human temperament and health.[50] In his *Esprit des lois*, Montesquieu was searching for the laws that govern human organization. He observed that the general character of people differs substantially in various climates. The second volume of Barthez' *Éléments de la science de l'homme* included a study of temperaments and an analysis of the effect of social and geographical factors on the development of man's physical and moral natures. He contended that climate, terrain, altitude, distance from the sea, winds, and the qualities of the sun affect human personality and behaviour. Even political factors alter the vital principle and consequently man's physical, mental, and moral make-up. Such studies were entirely consistent with the striving for an autonomous science of man to which the Société

[47] Ibid., vol. 1, pp. 91–121; vol. 2, pp. 319–328; Jean Cazeneuve, 'Philosophie de Cabanis', *Oeuvres philosophiques*, op. cit., note 40 above, vol. 1, pp. xxv–xxix; Picavet, op. cit., note 36 above, pp. 259–263.

[48] Jean Charles Grimaud, *Mémoire sur la nutrition*, Montpellier, Jean Martel, 1787, pp. 19–40; *Cours complet de physiologie*, 2 vols., 2nd ed., Paris, Egron, Gabon, Crochard, l'Heureux, Bechet et Lanthois, 1824, vol. 1, pp. 38–39; Xavier Bichat, *Recherches physiologiques sur la vie et la mort*, Paris, Brosson, Gabon, 1800, Art. 1, 1–9.

[49] This point was made by Frank E. Manuel, 'From equality to organicism', *J. Hist. Ideas*, 1956, **17**: 54–69. The difference between the notions of Condillac and of Cabanis led Aram Vartanian in the critical edition of la Mettrie's *L'homme machine*, Princeton University Press, 1960, pp. 125–129, to argue strenuously that, unlike Tracy, Cabanis was an opponent of Condillac's system. I believe that, though Cabanis moved a considerable distance from the older system, he did not break with it. Significantly, he remained in substantial agreement about the primary role of the sensations in the intellectual life, and he was an exponent of the system of analysis. Beyond this, however, the question is merely a quibble over categories.

[50] 'Traité des airs, des eaux et des lieux', in *Oeuvres complètes d'Hippocrate*, ed. and trans. by E. Littré, 10 vols., Paris, Baillière, 1840, vol. 2, pp. 12–93. Eng. trans. in G. E. R. Lloyd (editor), *Hippocratic writings*, Harmondsworth, Middx., Penguin Books, 1978, pp. 148–169.

Royale de Médecine and the Société d'Émulation had aspired. There was a commonly held conviction that morality, religion, the formation of societies, and so on would in time be shown to be on a par with physical laws and hence amenable to analysis.[51]

One of the most famous quotations in the history of medicine is that in which Cabanis described the brain as "a particular organ especially destined to produce [thought]; just as the stomach and the intestines operate the digestion; the liver filters the bile; the parotids and the maxillary and sublingual glands prepare the salivary juices." Sensory impressions entering the brain, he continued, induce it to action just as the sensation of food entering the stomach excites more abundant secretions of gastric juice. "The proper function of the one is to perceive each particular impression, to attach signs to it, to combine different impressions, to make comparisons between them, to take judgements and resolutions from them, just as the function of the other is to act on nutritive substances."[52] His more pious critics were outraged at what appeared to be an atheistic point of view. During the Bourbon Restoration, such statements provided a convenient excuse to dismiss and discredit Cabanis' work, causing it to fade into relative obscurity. It is incorrect, however, to accept so trivial and superficial an interpretation of Cabanis' words and thereby to dismiss him as a crude mechanist.

He was a materialist who, like La Mettrie and Diderot, proceeded from deeply rooted vitalist convictions. He shared their monist assumptions whereby the universe is a material unity.[53] Like them, Cabanis saw the world as a kind of nest of spontaneous, ubiquitous activity. Under certain conditions of organization, brute matter is transformed into living substance which is able to experience conscious or unconscious sensation. Matter possesses its own internal and spontaneous activity, which Cabanis attributed to a single universal force that is variously manifest in different forms of organization. It ranges from simple gravitation, to "electrical affinities", to "instinct of plants", and finally, to the complex nervous sensibility of animals.[54] The latter, he believed cannot be explained in terms of physical and chemical hypotheses but only in terms of the living body itself. One is reminded here of Barthez' hierarchy of natural motive forces ranging from attraction through to the forces of the vital principle.

Like all good scientific men by then, Cabanis ignored the question of first causes or "essences" in nature. He did not even refer to the soul as an intellectual principle. Nor did he make any reference to a creator God who transcended his creation. He is more accurately described as a pantheist or a deist than an atheist, for he believed that the

[51] Gusdorf, op. cit., note 3 above, pp. 180–212.

[52] Cabanis, op. cit., note 42 above, vol. 1, pp. 77–133.

[53] Ibid., pp. 263–293.

[54] See Martin S. Staum, *Medical components of Cabanis' science of man*, Princton University Press, 1981. Staum discusses Cabanis' debt to the animists and vitalists. He objects to the "materialist" label traditionally applied to Cabanis, who was by no means a mechanist reductionist. His phraseology resembled that of notorious atheists like d'Holbach and La Mettrie, says Staum, but he was very interested in the question of vital properties and human intelligence. See also Paul Delauney, 'L'évolution philosophique et médicale de biomechanicisme de Descartes à Boerhaave, de Leibniz à Cabanis', *Le progrès médicale*, 1927, **20**: 1289–1384; and Owsei Temkin, 'The philosophical background of Magendie's physiology', *Bull. Hist. Med.*, 1946, **20**: 10–35.

world is inhabited by an intelligent and providential will.[55] The human mind was presumed to participate in this universal intelligence. Such a monist notion wedded to the sensationalist belief that the development of a system of morals is based solely on responses to pleasure and to pain make reference to a deity unnecessary, at least for scientific purposes.

One of the clearest examples of the specific application of an analytic method to a medical question is the work on pathological classification done by *Philippe Pinel* (1745–1827), who exerted a major influence on Bichat. Pinel received a medical degree from Toulouse in 1773 before going to Montpellier. In 1777, being too poor to hire transport, he walked to Paris. There, he became a friend of Lavoisier and so came under the general influence of the ideologues. He began to study Locke and Condillac. Cabanis introduced him to the salon of Madame Helvétius. In 1792, on the recommendation of the Hospital Commission on which Cabanis sat, the painfully shy Pinel was appointed to take charge of the insane patients at the Hôpital Générale, established in 1656 to clear Paris of beggars. Its two major divisions were the Bicêtre, which confined males, and the Saltpêtrière, which was primarily for women. His much publicized move of releasing the patients, veritable prisoners, from their chains is reported to have saved his life on one occasion, when rumours that doctors were poisoning wells led to Pinel's seizure by a mob. He was rescued by a member of the French Guards, a former inmate who had been unchained.

As a *habitué* of the ideologues' groups, Pinel imbibed the arguments concerning the need to develop an autonomous science of medicine. It was assumed that the reform would include development of a rational and systematic nosology, or classification and description of disease. Vicq d'Azyr and other members of the Société Royale de Médecine had emphasized the need for such. As if to fill the need, Pinel's *Nosographie philosophique* appeared in 1798.[56] To do the work, he methodically applied the analytic method to decompose complex pathological states with a view to isolating the elements common to specific morbid phenomena. Mere observation of illness would not have been sufficient. According to Pinel, such classifications as those of Sauvages and William Cullen were too laborious, relying overmuch on such systems as Linnaeus' botanical classification. They failed to sort out the fundamental symptoms of a disease from its multitude of individual variations. Because there was frequent confusion between basic and secondary affections, there was an excessive and confusing multiplication of symptoms and illnesses. Pinel assumed that an illness was a sort of entity with a certain degree of regularity, a constancy of cause, effect, and consequences. By means of analysis, he intended to sort out and to isolate essential component parts of a disease from accidental individual symptoms.[57]

In a section on 'Method', Pinel advised the nosographer and the physician that the first step in studying an illness is to note the age and constitution of the individual, to retrace the disease's symptoms and to note the pulse, temperature, respiration, diges-

[55] See Cazeneuve, loc. cit., note 47 above, pp. xxxv–xxxvii.

[56] Pinel's *Nosographie philosophique* and its relationship to earlier classification systems and to ideology are discussed by Moravia, op. cit., note 2 above.

[57] Philippe Pinel, 'Introduction', *Nosographie philosophique ou la méthode de l'analyse appliquée à la médecine*, 2 vols., Paris, Richard, Caille et Ravier, 1798, vol. 1, pp. iv–xxxix.

tion, intellectual faculties, and excretions. One must not ignore such extenuating circumstances as fatigue, moral affections, abuse of pleasure, unhealthy air, contagious principles, and so on. Symptoms must be examined "independently of all hypotheses and only in terms of impressions made in the sensory organs." The second step requires the observer to separate individual variations of an illness from its specific characteristics. Finally, one must consider particular affections in relation to age, sex, manner of living, and habits which might modify the progress of an illness.[58]

Pinel divided disease into six classes: fevers, inflammations (*phlegmasies*), active haemorrhaging, nervous diseases, lymphatic affections, and "others". These were subdivided into orders, genera, and finally, a multitude of species. He consciously applied Condillac's method throughout. "It is necessary in the exposition as in the research for truth", he wrote "to begin with the most simple ideas and the ones which come immediately from the senses and then to move by degree to the most complex ideas."

Pinel arrived at six categories of fevers, which seemed to him to embrace all the known types: *angiotenic* fevers are characterized by an irritation of blood vessels; *meningogastric* fevers have their seat in the stomach membranes or in the membrane of the duodenum; *adenomeningeal* fevers shown an irritation of mucous membranes; *adynamic* fevers involve atony of muscle fibres; *ataxic* fevers follow a physical or moral shock to the nerves and finally, *adenonervous* fevers involve a contagious principle that affects nerves and glands.[59]

Inflammation, the second class of illness, includes a variety of contagious diseases. Ophthalmia, tonsillitis, various sorts of sore throat, and numerous catarrhs are all *inflammations of mucous membranes;* encephalitis, pleurisy, and peritonitis are *inflammations of diaphanous membranes;* "phlegmon", hepatitis, nephritis, and peripneumonia are *irritations of cellular tissue glands, and the visceral parenchyma;* rheumatism and gout are of the order of *muscle inflammations;* and finally, malignant pustules, skin inflammations, smallpox, measles, and scarlet fever are part of the order of *cutaneous inflammations.*[60] And so on with the remaining four classes of illness.

I list these categories only to make the point that in each case, the affected part was taken to be a particular fibre or member rather than a particular organ. *Angiotenic fever*, for example, does not attack the whole blood vessel but rather its tunic; inflammations are located in particular membranes rather than in the lungs, stomach, or other organs. It was an important innovation in pathological classification. It was also a catalyst for Bichat, leading him to differentiate the various membranes of the animal anatomy and to study their distribution and properties. In what was rather an uncharacteristic gesture for him, Bichat acknowledged Pinel to be the source of his ideas in the *Traité des membranes*, justifiably hastening to add that many of the

[58] Ibid., vol. 2, pp. 374–380; see also the 2nd ed., 3 vols., Paris, J. A. Brosson, 1803, vol. 3, pp. 523–530, where discussion is more extensive.

[59] Ibid., 1st ed., vol. 1, pp. 1–129, and 2nd ed., vol. 1, are devoted to a discussion of 'Fièvres'.

[60] Ibid., 1st ed., vol. 1, pp. 130–240, and 2nd ed., vol. 2, pp. 1–450, are devoted to a discussion of 'Phlegmasies' or inflammation.

specific results were entirely his own.[61]

Pinel, in turn, was quick to appreciate the importance of Bichat's theoretical innovation. In the second edition of the *Nosographie* in 1802, he commended Bichat's work calling it a most scrupulous and attentive examination of the subject.[62] By then, of course, the *Anatomie générale* had superseded the *Traité des membranes*. Although Pinel's first and second editions are the same for the most part, it is clear that his thoughts concerning his classification were affected by what Bichat had written. The implications of his own classification were thereby enhanced. Pinel wrote the following in 1802:

> The general view of all the theories of inflammation regard this term as unequivocal and as representing the same series of symptoms in all cases; whereas it may be taken with different senses depending upon whether it is located in the tegumens, the cellular tissue, the viscera, serous membranes, the articulation or the mucous membranes. But in these parts, so different amongst themselves, when one compares them for tissue, structure, sensibility and organic functions, do they not still have a common conformity in the lesions which they experience due to an irritating cause; and cannot one observe there the development, although in different degrees and in various proportions, the heat, the redness, the tension and the discomfort whose ensemble is indicated by the abstract terms inflammation?

The statement, which is not present in the first edition, indicates that Pinel had a fairly substantial involvement with the theory and the methods that Bichat used to study the tissues. Pinel wrote that he intended to combine the views of Haller, Bichat, and other modern physiologists to throw new light on the true character and the seat of cutaneous inflammations. He described the particularly effective method of isolating the epidermis by maceration. One can only detach the epidermis of a cadaver, he wrote, by means of putrefaction, maceration, or boiling.[63] This was precisely the sort of method which Bichat reported using to distinguish between membranes and then, subsequently, all the twenty-one tissues into which he divided the bodily parts.

The ideologues' range of interests was wide enough, as we have seen, to include all the natural sciences as well as man's political and social environment. During the

[61] Xavier Bichat, 'Dissertation sur les membranes, et sur leurs rapports généraux d'organisation', *Mémoires de la Société Médicale d'Émulation*, 1798, **2**: 371–385; and Xavier Bichat, *Traité des membranes en général et de diverses membranes en particulier*, Paris, Richard, Caille et Ravier, 1800, pp. 3–5. An instructive and interesting insight into Pinel's classification, and especially of its relationship with that of Bichat, is provided by Foucault, op. cit., note 27 above, esp. pp. 174–194. He makes the point that there is a great gulf between their two methods, for with Bichat, disease was no longer a particular entity but a complex movement of tissues in reaction to an irritating cause. For Pinel, fever was essential. For Bichat, it is a local irritation, an increase in the flow of blood accompanied by circulation.

[62] Pinel, op. cit., note 58 above, 2nd ed., vol. 1, pp. i–lvi, esp. xxxiii–xxxiv.

[63] Ibid., pp. 17–19. The connexion between Pinel's and Bichat's classification is even more clearly stated in a quotation in the 5th ed. of *Nosographie philosophique*, which is quoted by Pedro Lain Entralgo, 'Sensualism and vitalism in Bichat's "Anatomie Générale" ', *J. Hist. Med.*, 1948, **3**: 57–58: "The inflammatory state has, in fact, properties in common whatever the part attacked may be, and these points of contact are more considerable in proportion to the closeness of the analogy of the tissues and the organic functions of the parts affected; but again, how great is the diversity, if the organization of these parts is different! The phlegmasies have thus been divided into different orders, according as they are found in the integuments, in the mucous membranes, serous, fibrous, in the glands, or in the muscles. What does it matter that the arachnoid membrane, the pleura and the peritoneum are situated in the different parts of the human body, if these membranes possess general conformities in their structure?" With this quotation, Lain Entralgo implied that Pinel had a distinctly articulated tissue theory before Bichat published his work on membranes and tissues. Unfortunately, this statement is not present in the first or even the second edition of *Nosographie philosophique*. I believe that it reinforces the point that Bichat and Pinel mutually profited from each other's ideas.

eighteenth century, what were considered to be Newtonian scientific principles had been extended by a variety of enthusiasts to embrace the study of the human mind, society, and even theology. The ideologues had a hand in such developments. Because Napoleon suppressed the ideologues' teachings, however, the influence of sensationalism and analysis in the nineteenth century was only indirect. Certain deeply rooted notions concerning the nature of scientific investigation necessarily became implicit and thus remained part of the stock-in-trade of the scientific community. Temkin has shown, for example, that François Magendie was firmly rooted in the sensationalist tradition and in the vitalist assumptions that accompanied it.[64] Furthermore, the social sciences took form in the last century, building upon earlier foundations. Thanks at least partly to the preoccupations of the ideologues, sociology, political science, and above all psychology now had a rationale for their status as "sciences". This ostensibly put them on a par with physics, chemistry, and physiology. The names of Condillac, Destutt de Tracy, and Cabanis, however, largely vanished from the literature of intellectual history only to be restored in recent years.

We shall see in the next three chapters that sensationalist assumptions played a major role in Bichat's work. He belonged to the ideologues at least peripherally. Without being particularly careful to relate his ideas to those of anyone other than Pinel, Bichat described learning as a process of acquiring and relating sensations. The very idea of two lives upon which the arguments of *La vie et la mort* rested was possible only if one accepted such an epistemology. His much-lauded *Anatomie générale* is a classic example of a deliberate application of the method of analysis. Bichat decomposed the complex body parts and organs so as to isolate and to study the simple tissues composing them, and only his untimely death interrupted the recomposition he was undertaking in the *Anatomie descriptive*.

[64] Temkin, op. cit., note 54 above, pp. 20–35.

V

HOW TO STUDY LIFE: BICHAT'S METHOD

Although Bichat's only formal training was in surgery, his mature work, based upon a deliberate decision he made after Desault's death, was in anatomy and physiology. One of his most influential works, the *Recherches physiologiques sur la vie et la mort*, addressed the fundamental and timeless questions of the nature of life, the nature of death, and the ways in which these conditions of existence manifest themselves in an organism. The arguments developed in the book and the observations that reinforced them, were rooted in the assumptions of vitalism. Often, the experiments recounted appear haphazard and unfocused, a kind of "what if" approach rather than a deliberate attempt to address a question. It is unlikely that any of them could have jeopardized any of his a priori assumptions. The four-volume *Anatomie générale* was more satisfactory in this respect. It seems to have relied less upon a priori notions and more upon careful observations designed to analyse the tissues which compose the organs and other bodily structures. It is also the book whose reputation has best survived the years since its publication.

Some of Bichat's detractors have charged that his work is merely synthetic, and contains nothing original at all. What justification there is for the claim stems from the fact that Bichat dealt extensively with themes common to the eighteenth century. In the following chapters, Bichat's work will be analysed in relation to the theoretical and epistemological views already considered. Nevertheless, Bichat was much more than a mere borrower. His fellow-physicians, his students, and his successors at the Hôtel-Dieu invariably described him as a tireless worker who performed a great many post-mortem dissections as well as innumerable experiments on living animals. In this chapter, I wish to examine the method by which Bichat approached his work as well as some of its underlying assumptions.

The premiss basic to much of Bichat's work, as we have already seen, was that the life sciences are utterly separate from those that treat non-living matter. He developed the arguments in support of that thesis in the first part of *La vie et la mort*, his most complete statement of physiology. He presented evidence there in support of a natural division of living phenomena into animal and organic categories and described five vital properties that derive from sensibility and contractility. By so doing, he took these phenomena from the domain of mere sensation and locomotion and elevated them to be the physiological causes of formation, growth, and nutrition. We are in a position to recognize here an amalgam of notions gleaned from the vitalists, the monists, and the sensationalists. Naturally, he was never without evidence to support his contentions. Looking at Bichat's arguments in support of the animal-organic division of life, we see him writing as the heir to a tradition of interpretation that had its distant roots in the work of Aristotle and its more recent ones in that of Barthez and Grimaud.

The Montpellier tradition from the mid-eighteenth century maintained that observation is the sole reliable source of data about living creatures. This viewpoint was disseminated by its various spokesmen in the *Encyclopédie* and it was taken as belonging to vitalism, which emphasized the spontaneity of life. Claude Bernard cast Bichat into the mould also. In his eulogy of François Magendie, for example, he contrasted the experimental approach of his mentor with that of Bichat, claiming that in his work, "the broad philosophical views of Bordeu subdued and killed the experimental method of Haller". Yet Bichat's own work demonstrates that he himself did not consider vitalism to be a conceptual barrier to experimentation. Indeed, he tried consciously to reconcile the Montpellier approach with that of experiment, asserting that his goal was to ally "the experimental view of Haller and Spallanzani with the broad philosophical views of Bordeu".[1] He made little distinction between experiment and observation. In any case, the variability of the life forces precluded only the mathematization of physiology, not experimentation. So it is that the first part of *La vie et la mort* is largely constructed of arguments having to do with observations of living phenomena. In the second part, however, the phenomena of death are tested by experiment.

Indeed, Bichat approached this second part quite differently from the first. Having dealt earlier with the gradual death of the body due to ageing, Bichat then undertook to examine what happens in violent, accidental, and sudden death. He concentrated on the role of the brain, the heart, and the lungs, for he believed that all sudden death commences with the interruption of either the circulation, the action of the brain, or the respiration. His rationale for such an undertaking was rooted in the notion of the two lives. The primary bodily functions, he believed, proceed from the heart, which is the major organ of the organic life; the brain, which is the centre of the animal life; and the lungs, which participate in both lives and mediate between them. The action of each of these organs, the three "centres" of life, is necessary for the other two.[2] The experiments and observations recorded here supplemented the speculations of the first part of *La vie et la mort*.

Bichat's basic procedure was simply to provoke injuries in specific organs of various animals and to observe the consequences of such actions for the rest of the body. He used animals in abundance. He also frequently attended executions by guillotine so as to be able to make observations on the severed heads and trunks of its victims. After examining the effects of heart injury, Bichat concluded that red blood influences the brain directly by its motion rather than by means of any principles it carries. His initial arguments in support of this notion were somewhat tortuous inferences from anatomical design. He wrote, for example, that animals with long necks and hence with brains rather distant from their hearts appear to be less intelligent than those with short necks. It also appeared to him that the great arterial trunks located at the base of the brain were designed precisely to facilitate the brain's receipt of vascular motion. Finally, the brain's bony home, unlike that of most other organs that are embedded in soft cellular tissue, appeared to him to be specifically designed to reflect heartbeats.

[1] The question of experiment in relation to physiological theory is discussed by William Randall Albury, 'Experiment and explanation in the physiology of Bichat and Magendie', *Stud. Hist. Biol.*, 1977, **1**: 47–131. The quote from Claude Bernard's *Éloge de Magendie* is taken from this article.

[2] Xavier Bichat, *Recherches physiologiques sur la vie et la mort*, Paris, Brosson, Gabon, 1800, pp. 191–196.

Some readers in 1800 would certainly have found such arguments to be as shallow and unpersuasive as we do. They were, however, merely offered as a preliminary to experimental evidence, which Bichat believed pointed toward the same conclusions. He found, for example, that water injected slowly into an animal's carotid artery had little or no effect, whereas that injected more quickly produced a troubled cerebral activity or even death. An arterial haemorrhage produced a gradual loss of brain-induced activity. Air injected into the veins or arteries in any quantity frequently produced brain injury or death, suggesting to Bichat that it cushions the heartbeat, damping the transmission of motion to the brain.

His conclusion about the blood pounding the brain led Bichat into a long chain of inferences concerning the effect of heart injury upon general bodily processes. Although the influence of the left heart on the brain is through the blood vessels, Bichat believed that that of the brain on the heart and other organs must be through the nerves. It had long been known that an interruption of brain activity paralyses certain nerves and through them, the intercostal muscles and the diaphragm. Because it is centred on the brain, Bichat reasoned that the animal life ceases the moment excitation from that organ is interrupted. Being strangers to the brain's direct influence, however, the organs of the organic life fail largely as a result of the preliminary failure of the circulation, which no longer transports to them the materials they need in order to function. The end of nutrition, exhalation, secretion, and digestion, therefore, follows only gradually. In general, then, Bichat's conclusions about the death of the heart owed less to experiment than to deductions based on certain assumptions concerning the function of the heart and the brain.[3]

Bichat's second investigation, having to do with the consequence of injury or death to the lungs, is, in many ways, the most interesting part of this investigation. The physiological role of respiration was a timely question. Much light had been shed upon it in the preceding century or so by extensive chemical studies of gases. By Bichat's time, oxygen had been identified as the crucial respiratory element, while fixed air or carbonic acid gas, which we call carbon dioxide, was known to be the waste gas discarded in expiration. United with oxygen, blood acquires a crimson hue; united with fixed air, it is dark or black. Red blood carries life and vitality to the parts through which it flows, while black blood transports weakness and ultimately death in the form of asphyxia. By the 1790s, any physiologist had these facts to draw upon.

Bichat distinguished between mechanical and chemical interruption of the activity of the lungs, even though the one condition inevitably produces the other. He could cause mechanical failure by, for example, opening the chests of animals, thereby immobilizing the lungs by the force of external air pressure. Chemical failure was produced by obstructing the trachea, creating a vacuum around an animal, or plung-

[3] Ibid., pp. 197–238. A very recent and interesting paper on this subject is Geoffrey Sutton, 'The physical and chemical path to vitalism: Xavier Bichat's *Physiological researches on life and death*', *Bull. Hist. Med.*, 1984, **58**: 53–71. Sutton points out that in these experiments, Bichat set out to show that the heart has the primary function of sustaining the tissues of the brain by agitating them. The heart's fundamental role in the body is not only to distribute nourishment but by its motion to sustain all the tissues. At bottom, Sutton claims, the discussion offered a disquisition on the importance of mechanical agitation of the sensible organic system, in the form of pulsations, oscillations, and shocks.

ing it into any number of gases. His subjects included a somewhat arbitrary mixture of hanged criminals, animals that he had strangled, or that he had drowned slowly in stages, and so on. Although it is impossible to be precise, his victims were easily numbered in dozens for this section of his work alone.

The various procedures permitted him a great deal of control over the type and quantity of air available to the body. One of his preoccupations came to be to observe the way in which blood and the various organs are affected by a lung injury, or as the result of altering the availability of air or its condition. He examined pregnant dogs and guinea pigs and found that the umbilical vessels and the foetus are affected as rapidly as the mother's own circulation. Asphyxia, he discovered, is more rapid with nitrous or hydrogen sulphide gas than with carbon dioxide, nitrogen, hydrogen, water, or a vacuum. He even tried to note, with no particular success, the differences in response caused by such factors as age and temperament. Finally, Bichat observed that the close connexion visible between the functions of the lungs and those of the heart in warm-blooded animals does not exist in cold-blooded ones.

In every case, Bichat found that it is the black blood that enfeebles the tissues of the body, accounting for its gradual weakening. Does the black blood act upon the fibres themselves or upon the nerves? Although Bichat inclined toward the latter opinion and thought it likely that oxygen is the principle of irritability that activates the tissues, he reminded himself somewhat pompously here of the limits of scientific observation. "Let us stop", he wrote, "when we arrive at the limits of rigorous observation; let us not strive to penetrate there where experience cannot clear the way for us." This attitude had to do, of course, with the frequently stated reluctance of the eighteenth-century scientist to shape hypotheses going beyond the capacity of his data to test. It was a tribute to the Newtonian notion of limited explanation and was born of a disdain for an intellectual exercise that quested for the elusive "first causes" of science. It was also a basic tenet for sensationalists, whose notions of epistemology and explanation Bichat accepted. By quibbling over the evidence supporting the respiratory role of oxygen, however, Bichat revealed, above all, that his knowledge of chemistry was extremely limited. We shall shortly come across a manifestation of this same basic attitude in connexion with Bichat's tissue work, where his sensationalist principles were interpreted as proscribing the use of microscopic observations in anatomy.

Bichat's surgical skill was to stand him in good stead in some rather complicated experiments that he devised to examine the effects of asphyxia on brain functions. He had often observed, he wrote, that the arterial blood of one animal transfused into the carotid artery of another had no effect upon brain function. The venous blood of one animal, on the other hand, though it produces varying results, always causes eventual death when it is diverted to bathe the brain of another animal. That death, he concluded, must be the consequence of asphyxia produced by black blood. As long as the organic life has not yet been extinguished, the asphyxia can be reversed by reintroducing red blood. Again, when he looked for the same phenomena in such cold-blooded animals as frogs and fish, the results were much less clear.[4]

The violence and apparently crude indifference of medical investigators in the

[4] Bichat, op. cit., note 2 above, pp. 239–369.

eighteenth century toward animals must impress even the most insensitive contemporary reader of their texts. Even if one admits that the infliction of some pain was unavoidable if they were to do their work, the lack of compassion is really quite remarkable. I know of no investigator besides the unhappy and opium-addicted Haller who ever expressed any revulsion at the enormous price exacted from animals. The kindliest researchers and teachers, affectionately remembered by their students, colleagues, and families, seem never to have flinched at the prospect of performing the most hideous experiment, no matter how trivial the anticipated result might be. The mild-mannered Bichat was no exception. For example, he blithely tossed beasts into fire, water, and so on to try to determine in some vague way, and for some ill-defined reason, the different effects that follow from the various types of asphyxiation. The results seem to have been of minimal significance, if any. The best he could have hoped for was a kind of inductive accumulation of data that might possibly at some time be incorporated into a larger theory of organic function. For example, for some obscure reason, Bichat killed a considerable number of animals by injecting such fluids as ink, oil, wine, coloured water, urine, bile, and mucous fluids into their arteries. His only conclusion was the dubious one that these various substances acted on the brain rather than on the arterial surface, because they were more frequently fatal when they were injected into the carotid rather than into, say, the crural artery. Even in his most optimistic and euphoric moments, he could not have anticipated any but minimally important results. At worst, there was virtually no coherent point at all to such experiments.

The final series of experiments in *La vie et la mort* were concerned with the question of the consequences of brain injury. Bichat observed many times that section of the part of the vagus nerve connecting the brain and the lungs will not arrest respiration. On the other hand, section of the spinal marrow between the last cervical and the first dorsal vertebra does so by paralysing the diaphragm and the intercostal muscles, which he claimed belong to the animal life. This confirmed Bichat in his belief that respiration is a mixed animal-organic function. That is to say, he claimed that it belongs to the animal life because it involves voluntary muscular motion and to the organic life because its chemical functions are linked to the heart.

In animals and in guillotined human bodies, Bichat frequently observed that the heart may be made to contract by direct stimulation for a considerable time after death. Stimulation of the cardiac nerves, vagi, or medulla spinalis, however, has no effect. He took this to be additional evidence for the independence of the heart from the brain and for the separation of the organic life from the animal. The heart stops after injury to the brain, he claimed, only because of the cessation of the activity of those intermediate lungs. Virtually all his observations and experiments, therefore, were somehow related to the animal-organic division, and all of them apparently confirmed it as genuine and natural. Having once satisfied himself that it existed, it never seems to have occurred to Bichat that his division was untested or merely a classificatory convenience. He did not, therefore, analyse it.[5]

Bichat's other great work was his *Anatomie générale*. It, and the *Traité des*

[5] Ibid., pp. 370–434.

membranes, which preceded it, were based deliberately upon techniques of analysis that were much emphasized at the time. In the *Anatomie générale*, Bichat argued that living matter is compounded from twenty-one basic elements, which he called the tissues. That is, Bichat undertook to break such complex structures as organs into their simpler component parts and to study these parts separately. At least in theory, Bichat decomposed some bodily part until it was no longer amenable to further breakdown by any of a number of techniques. He then declared that smallest unit to be a tissue and proceeded to determine its particular distribution of physical and vital properties. He believed that sensibility and contractility, the vital forces, belong to the elemental tissues and through them govern the integrated activity of the body.

The tissue theory was an extension of Bichat's membrane work. His 1798 paper, 'Mémoire sur la membrane synoviale des articulations', was his first offering on the theme. Although Bordeu's name is not even mentioned in it, it was clearly modelled on his study of the glands, and reads as though that work were at Bichat's elbow as he wrote. Using words unmistakably like those of Bordeu, Bichat opened the paper with the assertion that no part of physiology is richer in hypotheses but poorer in facts. His main purpose in doing the work, he continued, was to demonstrate the inadequacy of the mechanist theories that had been adopted thus far to explain how synovial fiuid is carried to an articular surface. The synovial system of the body extrudes a viscous fluid which lubricates the joints and tendon sheaths. This fluid cannot be a glandular product, simply because painstaking examination showed that no glands exist in the regions in question. Nor does the marrow extrude the fluid, as he found that its destruction is without effect. The only remaining possibility was that synovial fluid is produced by an active, that is to say, vital, process of exhalation. Bordeu had concluded that each gland in the body secretes its own particular humour because it possesses a unique sensibility. Like him, Bichat remarked that each class of organs had "its proper life independent of that of the other classes of organs".[6]

That paper was meant to be read along with an accompanying 'Dissertation sur les membranes', which itself was an important innovation in the manner of viewing the bodily parts. Bichat wrote that membranes tended merely to be associated with the organs over which they are spread. The pericardium and the heart, or the pleura and the lungs, for example, were always considered together. Although Haller had treated membranes as unique structures, he had not differentiated between them, for he believed them to be merely structural modifications of the organs. The first entirely new insight had been that of Pinel in his *Nosographie philosophique*. Pinel had established "a judicious connexion between the different structures and the different affections of membranes; . . . ."[7] As we have seen, Pinel observed that inflammations, for example, may variously affect the cutaneous surface, cellular tissue, glands, serous membranes, muscles and articulations, and mucous membranes. He classified the dysfunction accordingly. From reading Pinel's work, Bichat formed the idea that membranes of the body can be analysed into the mucous, fibrous, and serous varieties

[6] Xavier Bichat, 'Mémoire sur la membrane synoviale des articulations', *Mémoires de la Société Médicale d'Emulation*, 1798, **2**: 351–370.

[7] Xavier Bichat, *Traité des membranes*, Paris, Richard, Caille et Ravier, 1800, pp. 3–5.

and into compounds of these basic tissues.[8]

At this stage, in 1798, Bichat's method was largely one of painstaking dissection and observation. With it, he established that mucous membranes, so named because of the fluid secreted by glands just beneath their surface, line all the hollow organs that communicate with the exterior of the body, forming the gastrointestinal tract, bladder, womb, nasal fossae, and all excretory ducts. The moist, highly polished serous membranes, he observed further, are generally found in the form of envelopes around those organs whose interiors are lined by mucous membranes. They occur around such contractile organs as the heart, lungs, stomach, intestines, and womb, in which organs they form the pericardium, pleura, peritoneum, and vaginal tunic. Fibrous membranes, not moistened by any fluid, are especially numerous, forming the periosteum, which covers bones; the dura mater of the central nervous system, the sclerotica of the eye; the envelope of the kidney; the internal tunic of the spleen; and the albuginea of the testicles. They form the aponeuroses, which encircle the thigh, leg, arm, and forearm, and the fibrous capsules that cover the synovial membranes of the articulations in the femur and the humerus. Tendons and ligaments are composed of the same strong, elastic, and insensible matter. Finally, Bichat's elaborate dissections showed him that these membranes send numerous extensions into organs, thereby producing a kind of fibrous skeleton that sustains the tissues of testicles, kidneys, and spleen.

There are also "composite membranes", Bichat wrote, which adhere to one another in such a way that a single membrane is produced by the union of two other types. By the time he composed the *Traité des membranes*, Bichat found that he was encountering membranes that did not belong to any of those six subdivisions. The middle tunic of arteries is normally classed among the muscular organs, but Bichat thought it demonstrated certain characteristics of fibrous membranes. The inner tunic of blood vessels was different in veins and arteries. There were such unknown membranes as the lining of the medullary canal of bones, the iris, the choroides, the retina, and the pia mater. In certain cases, the body forms unnatural membranes such as cysts, cicatrices, and scars.[9]

In the original paper, Bichat had little to say about the vital forces in various membranes. By the time of the *Traité des membranes*, he had come to consider their distribution to be as characteristic of a membrane as its anatomical properties. He searched for sensibility, just as Haller had done half a century before. Mucous membranes, Bichat found, possess a "lively sensibility" because of structures resembling papillary bodies located beneath the surface of the epidermis. Serous membranes normally possess only an obscure sensibility. Nature has so designed them because ordinarily they are not in contact with foreign bodies. Haller had classified fibrous structures among insensible organs, but Bichat found that they become painful when

[8] Othmar Keel, 'Les conditions de la décomposition "Analytique" de l'organisme: Haller, Hunter, Bichat', *Études Philosophiques*, 1982, no. 1, 37–62, argues that Bichat's claim of indebtedness to Pinel notwithstanding, the method of analysis or decomposition was present long before the time of the ideologues in the work of Haller, Bonn, Hunter, and others.

[9] Xavier Bichat, 'Dissertation sur les membranes, et sur leurs rapports généraux d'organisation', *Mémoires de la Société Médicale d'Émulation*, 1798, **2**: 371–385.

exposed to air for a time. While such fibrous structures as ligaments are not irritated by chemical agents, they are painfully affected by agents that cut, tear, or dislocate them. Like Whytt, Bichat remarked that one should never pronounce upon the insensibility of any organ until one has exhausted all means of irritating it.

With his *Anatomie générale*, Bichat took this work a very significant stage further. By incorporating into it all the material found in the *Traité des membranes*, he made the earlier work redundant. In the fifteen months separating the publication of the two works, he had analysed all the solid parts of the body into elements and observed and recorded their various vital and physical properties and responses. The membranes became but four of twenty-one tissues. The theories of the two lives, the vital properties, and properties of texture were wedded to the notion of physiological elements to produce the tissue theory, which was at least a precursor to histology.

Bichat proceeded in an orderly and thorough fashion, providing as much information as possible about the structure, distribution, properties, and particular functions of every one of the tissues he had identified. A brief glance at his extensive analysis of cellular tissue, the most abundant of all the parts of the body, should illustrate Bichat's general approach to the living elements.[10] He described cellular tissue first as it appears when one is merely tracing its distribution during the course of a dissection. He found it lining one side of the skin and the serous and mucous membranes and external to the arteries, veins, absorbents, and excretories. The quantity of the tissue, its form, and the strength of its adherence to the structures it accompanies were discussed in every case. Sometimes, even when the presence of the tissue was not obvious, it could nevertheless be shown to exist by indirect methods.

> In many of these organs . . . it may be made perfectly distinct by maceration, which insensibly softens and separates their fibres, as in the tendons and fibrous membranes. Boiling deprives some of their nutritious matter, gelatine for instance, and leaves a membranous residue which is evidently cellular. In all, even bones, cartilage and so on, granulation, the production of which, as we shall find, is specifically of a cellular nature, proves the existence of this internal tissue, of which there are so many processes.[11]

His observations convinced Bichat that, as well as covering and thereby insulating the organs from one another, cellular tissue formed one of the principal elements uniting their parts.

Bichat next outlined for his readers the distribution of his system throughout the various parts of the skull, face, neck, chest, abdomen, pelvis, and the extremities. He described the variously shaped "cells" or gaps that gave the system its name and that exist to absorb fat or fluids as necessary. Examining the serum of the cellular tissue, he found that it varies in quantity, depending on where it is in the body. It is most abundant in such parts as the scrotum, eyelids, and prepuce, which are deprived of natural fat. Chemical experiments show it to be albuminous in nature. He described the distribution of cellular fat throughout the various organs, remarking also upon its various states in health, disease, and age. He speculated about how it is separated from the blood during the course of its formation, without, however, arriving at a conclusion.

[10] Xavier Bichat, *Anatomie générale appliquée à la physiologie et à la médecine*, 4 vols., Paris, Brosson, Gabon, 1801, vol. 1, pp. 11–114.

[11] Ibid., p. 32.

Under the heading 'Composition of cellular tissue', Bichat described the behaviour of the soft and delicate cellular tissue in response to various substances and reagents. When exposed to air, cellular tissue dries quickly and remains white. It does not putrefy as rapidly as, for example, the glandular or muscular parts. He found it to be considerably resistant to soaking in water and even to boiling. Forcing himself to vomit food he had ingested earlier, he determined that gastric juices affect cellular tissue less than they do the fleshy or muscular parts of the body. This type of analysis illustrates well Bichat's basic assumption concerning the elemental nature of a tissue. An isolated chemical element, by definition, behaves consistently in response to any treatment. And so it is with bodily elements. Because of its unique nature, cellular substance will behave consistently in response to various types of physical or chemical treatment.

Very significantly for his general purposes, Bichat examined the tissue and vital properties. He attributed extensibility to cellular tissue simply because of its striking ability to expand in response to oedema, fat accumulation, tumours, and so on. It must be extensible to permit the motion of the limbs and the flexion and extension of every part. It follows that it must also possess contractility, simply because the tissue contracts whenever extensibility ceases.

Searching for what he described as the "animal properties", Bichat sounded unmistakably like Haller:

> You may divide it as you will with the scalpel, draw it out in every direction and distend it with gases. The animal when submitted to these experiments shows no signs of pain; if any be felt, it is produced from the threads of nervous tissue which supply it, and may perchance have been accidentally injured. In a morbid state, on the contrary, the sensibility of the part is increased to that degree, that it becomes the seat of the most acute pains. We have an obvious instance of this in phlegmonous inflammation.[12]

He had had little to say about the "organic properties" of the membranes earlier. Now, however, he directed his attention to them as much as to the animal properties. He could only infer their presence in the tissues from the functions that they were alleged to perform, because, by definition, organic sensibility and contractility are neither consciously regulated nor perceived. He assumed that organic properties are present in every living tissue, because, according to the physiological system worked out in *La vie et la mort*, he had assigned to them the most fundamental functions without which life could not exist. He assumed, therefore, that cellular structures must have the level of organic sensibility and contractility necessary to maintain their nutrition. In addition, the cellular tissue must have a very specific type and quantity of the organic forces to account for its absorption of fat and serum. "The existence of insensible organic contractility is indisputably proved", he wrote, "in the cellular tissue by the processes of exhalation and absorption taking place there." Because his clinical experience had taught him that suppuration and inflammation occur frequently in cellular tissue, he wrote, "it is obvious that the principle of life is abundantly developed in the cellular system. . . . The phenomena of inflammation for this reason run their career with great celerity in this system."[13]

Bichat remarked upon the fact that, unlike most other organs, cellular tissue

[12] Ibid., p. 81.
[13] Ibid., p. 82.

possesses the faculty of extending and reproducing itself. The formation of cysts, cicatrices, and tumours all depend on the ability of the cellular tissue, which is their common base, to reproduce itself. He observed the development of each one of these formations and outlined his observations in considerable detail.

Finally, Bichat examined the development of cellular tissue, looking for its presence shortly after conception, in an older foetus, and in the various stages of life. In some cases, he looked specifically for the tissue in animals that he dissected. In others, however, his information simply came from observations made upon the large number of subjects he personally dissected or whose dissection he supervised in the course of his clinical work.

Bichat's approach to cellular tissue has been outlined in rather tedious detail simply because it is typical of his approach to all the tissues. In general, his technique may be summed up as one involving a combination of thorough dissection and exploration of the body as a whole in various conditions of age and health with an examination of the tissues under the influence of various physical and chemical reagents and a search for their vital and tissues properties. His search for the animal forces of life, as we have seen, involved the use of the method Haller had described in his work on sensibility and contractility. His search for the organic forces, on the other hand, was altogether indirect, having to do primarily with an observation of the functions that the tissue performs. If, like the cellular or glandular tissue, it performs extensive secretory and absorptive functions, Bichat inferred that it must be rich in organic properties. Inevitably, he was merely extrapolating from what he assumed he had clearly established in the course of formulating his physiological theory.

Bichat's treatment of the exhalant tissues of the body is particularly interesting, simply because no such structure is visible to the naked eye. Bichat devoted many pages in the *Anatomie générale* to a discussion of the distinction between excretion and exhalation, which he had first raised in his 1798 paper on the synovial membrane. Most authors, he claimed, tended to call everything secretion. But, "On being guided by inspection only and without penetrating to enter into the intimate nature of the organ, it is evident that whenever secretion occurs, a gland exists." Exhalations, on the other hand, which are deposited over the serous, mucous, synovial, cutaneous, and cellular surfaces, are extracted from the circulating fluids directly without the intermediary of a gland. At least stretching, if not breaking with, his own sensationalist rules concerning clear and simple demonstration, he stated that he would proceed by "strict reasoning" in his study of these largely hypothetical "tissues". After a discussion of why the invisible exhalants must be present in the body, he wrote: "All duly considered, 1st. The existence of exhalants, 2ndly. Their origin from the capillary system of the part where they are found and 3rdly. Their termination on diverse surfaces are the only facts correctly ascertained."[14] But he was merely guessing, of course. It is particularly ironic that he should have done so in view of his frequently voiced aversion to anything but clear and simple sensory demonstration. It was Bichat, after all, as an earlier quotation shows, who had quibbled over the reliability of the evidence in support of the theory that oxygen is the essential respiratory gas.

Even more curious is the fact that Bichat refused to use a microscope in his

[14] Ibid., vol. 2, pp. 549–576.

investigations and that he denied any validity to the evidence it provided. It might conceivably have confirmed the existence of the exhalant tissues, but Bichat would not have considered using it to look for them. This stance has aroused curiosity and, occasionally, acrimony ever since. It seems particularly perverse of the man who has been dubbed the founder of histology to construct his theory all the while disdaining the use of what has since become its most important instrument. On the other hand, Bichat's contemporaries, at least in the Paris clinical school, understood his position very well. The instrument had, of course, made certain contributions to medical observation before 1800, but their importance is a matter for debate even today. In any case, certain observations could have been directly related to aspects of Bichat's work, or so it appears today with hindsight. For example, in 1660, Marcello Malpighi had used a microscope to observe the capillaries, whose existence William Harvey could merely postulate in 1628. By so doing, Malpighi had provided the last element necessary to complete the theory of blood circulation.

Because he would not observe them with a microscope, Bichat was unable to learn anything specific about the structure of these small, thin vessels. He believed in their existence nevertheless, devoting a section of the *Anatomie genérale* to them. What evidence we have, he claimed, is limited to what is observed during inflammation. He found also that injection of a coloured fluid into the fine artery of a cadaver will reveal that such vessels exist in every part of the body, but "Such is their tenuity that up to this point we have had no facts grounded in experience and observations."[15] Similarly, he dismissed the microscopic studies done by Leeuwenhoek and others on muscle fibres, claiming that such an examination of the intimate structure of organs is merely a futile search for inaccessible first causes "whose knowledge would add nothing to physiological notions on the motion of muscles".[16] Although the microscope in the hands of Malpighi and Ruysch had revealed much about glandular structure, Bichat dismissed it as pointless.[17]

Bichat's reluctance to use the microscope in the analysis of tissues can best be understood in relation to sensationalist notions concerning the proper study of nature. Microscopy implied a search for first causes, and as we have had occasion to observe, no eighteenth-century scientist grounded in sensationalist principles would admit to such an exercise. Bichat admonished his readers as follows:

> Let us neglect all these idle questions where neither inspection nor experience can guide us. Let us begin to study anatomy there where the organs begin to fall into the range of our senses. The rigorous progress of sciences in this century does not accommodate itself at all to these hypotheses which have been nothing but a frivolous fiction of general anatomy and physiology in the previous century.[18]

This statement is actually the crux of the matter, providing the clue to a prejudice that Bichat shared with many other medical men. John Locke and Thomas Sydenham, both judged to be outstanding medical pioneers by Bichat's contemporaries, had rejected the microscope, arguing that a search for the intimate material bases of disease contributes nothing to medical practice. David Wolfe considered this attitude to be a manifestation of Puritan morality. Subsequently, Laënnec, Cabanis, Pinel,

[15] Ibid., pp. 469–548 deal with the capillary system. The quotation is on p. 507.
[16] Ibid., pp. 224–338 deal with the muscular system of the animal life. The quotation is on p. 231.
[17] Ibid., vol. 4, pp. 569–639.
[18] Ibid., p. 576.

and other exponents of the "cult of observation" of the Paris school mistrusted the use of instruments, because they believed that only knowledge garnered from experience is legitimate. To sharpen the human senses artificially was to distort their role in the process of observation and to transgress a fundamental rule of procedure.[19]

The rejection of an instrument that we regard as having enormous potential importance for anatomy seems akin to the attitude of the apparently reactionary Aristotelians who, some two centuries earlier, had refused to look through Galileo's telescope. They could not be persuaded that the images that suddenly become visible when some lenses are interposed between one's eye and an object exist in reality. Neither they nor the sceptical seventeenth- and eighteenth-century physicians were being entirely unreasonable. Many beautiful microscopes by skilled artisans survive, demonstrating that they existed in large numbers. They tended, however, to be largely the playthings of educated amateurs. Genteel folk owned them, much as many people today own microscopes or telescopes simply because of a largely passive interest in the objects of nature. The microscope was more of a tool used to gaze upon such marvels of creation as an intricate butterfly's wing that one intended for purposes of discovery. Indeed, there was relatively little important scientific work done in the eighteenth century that owed much to the microscope. If physicians owned them, it was primarily because they were gentlemen rather than because they were medical men.

Leeuwenhoek's and Malpighi's work notwithstanding, Bichat and his equally sceptical fellow-physicians and researchers were probably right to remain suspicious of microscopic evidence. The value of the illustrations they had left behind was dubious to begin with. Leeuwenhoek, for example, had been very secretive about his handheld single-lens microscopes, not permitting even the artist whom he commissioned to do his illustrations to look through them. Apart from that, until about the 1830s, observations made with microscopes were seriously hampered by spherical and chromatic aberration. Even today, an untrained observer such as a student, as often as not, sees only what he has been told he should expect. With the cruder instruments and staining techniques that would have been available to Bichat and to his predecessors, the problem of interpreting a vague and complex image would have been far greater still. If we judge his scepticism as a streak of reaction, it is merely because with hindsight we believe that the microscopists were on the right track and their work would have reinforced Bichat's observations. He, on the other hand, was showing exemplary scientific caution in the face of a dubious instrument that was still little more than a rich man's toy.

It is doubtful whether Bichat's work on general anatomy would have been enriched even by the assiduous use of the microscope. Probably the only way to answer such a question would be to look at the sorts of tissue specimens he described with a microscope of the period. His prohibition against it, interestingly, did not extend to the hand glass, which he used without apology. Nor is this completely inconsistent, if we bear in mind that the hand glass is little more than a kind of spectacle lens, magnifying but slightly, distorting apparently not at all, and requiring no preparation of the specimens. It helps an observer to see rather than bringing new images before him for interpretation. The strictures against the dubious and mysterious effects of elaborate

[19] David Wolfe, 'Sydenham and Locke on the limits of anatomy', *Bull. Hist. Med.*, 1961, **35**: 193–220.

instruments need not, therefore, apply to it.[20]

Since Bichat disdained the use of instruments to extend his senses, it would have been consistent for him to avoid formulating theories that were untestable because they were also beyond the range of those same senses. For example, because microscopic observation of capillaries was prohibited, it should have been equally unacceptable to speculate about the existence and nature of organic sensibility and organic contractility, about the production of heat, and about the nature of inflammation in those same all but invisible capillaries. Whereas the properties he attributed to the animal life are easily observed in the body, the existence of the remaining vital activities was merely conjectured. They are, in a sense, as much microscopic properties as minute fibres are microscopic structures. Bichat's confidence in the existence of those alleged forces arose from the mere fact that every bodily part must be nourished, must exhale, absorb, and secrete. However justifiable such confidence seemed to him to be, it was not based on the evidence of his senses. He therefore transgressed his own sensationalist principles and rules of procedure.

In all of Bichat's work, there is not one single illustration. It must appear to any modern reader to be a major omission in a work of general anatomy. Indeed, it would often have been far easier to read his books if he had resorted to at least the odd diagram to supplement a complex verbal description of the form or the distribution of some structure. It would seem that having written often enough about the sufficiency of sensory evidence for the scientific investigator, he might have obliged his readers by showing them roughly what his own skilfully trained visual senses had focused upon. If he thought about it at all, Bichat might possibly have assumed that the type of person to whom he addressed the book, the physician and the medical student, would not have missed illustrations too sorely. It is also possible, however, that Bichat avoided illustration for more fundamental reasons. The *Anatomie générale*, after all, was about the tissues whose essential structure could no more be visualized than could the essential structure of Lavoisier's chemical elements. A multitude of his anatomist predecessors had extensively and adequately illustrated the objects of gross anatomy. But the anatomical elements with which Bichat was preoccupied were beyond the scientists' ability to picture. If Bichat felt so, he would not draw them for the same reason that he would not use a microscope. But that is merely speculation, for Bichat showed no sign that he felt his omission required explanation.

To have completed so vast a quantity of work in a few years, combining it with his clinical work at the Hôtel-Dieu, is impressive, and accounts for Bichat's reputation as a tireless worker. The remainder of my discussion will focus upon the way in which it fits into eighteenth-century currents of thought concerning the nature of life and of living matter. Though I shall write of borrowing and of synthesis, it should be clear by now that his ideas were equally derived from a large amount of observation and testing. The work of his predecessors was there, he knew it well, and he used it, as should become clear. Nevertheless, he did so always with a view to having it merely to assist him in confronting raw material that he had personally acquired

[20] I am indebted to David Bryden, former curator of the Whipple Museum of Cambridge University for useful insights and information concerning the nature and technical limitations of early microscopes.

VI

THE FORCES OF LIFE AND THE CAUSES OF DEATH

Xavier Bichat's *Recherches physiologiques sur la vie et la mort* was published the year that the century turned. This seems fitting, for the work is a superb synthesis of a great many of the important concepts concerning the living body which had evolved during the eighteenth century. In a sense, it sums up the state that medical theory had achieved by them. While studying *La vie et la mort*, one sees unmistakably that Bichat was an heir to the vitalists, the organicists, and the sensationalists. In it, Bichat examined the living organism with a view to discerning and describing those characteristics which he believed divide it so utterly from the inert one. Here, he argued with enormous conviction on behalf of the separation of the science of life from that of non-living things with which, he believed, it shares virtually no common ground. Here also, he developed his arguments on behalf of subdividing physiological forces and processes into animal and organic categories. Each life was examined in terms of its supposed properties, its consciousness, its ability to learn, its birth, and its death.

Had Bichat not gone to Paris in 1793, it is unlikely that he would have been able to achieve this and hence any other of his major works. Like a great magnet, that city has always drawn France's most ambitious and creative men to itself. Even in the 1790s, the decade of the most intense travail it has ever suffered, it held out to Bichat and to others the possibility of an opportunity to study, to make important contacts, and to achieve professional advancement in general. As we have seen, Bichat obtained all these things from Paris, largely as a consequence of his very lucky adoption by Desault. The vitalist theory was, by then, widely diffused. And the forces of sensibility and contractility, which preoccupied the organicists, had been, as we have seen, the subjects of assiduous examination for about half a century before Bichat tackled them. The sensationalist theory, albeit of at least seventeenth-century origin, was being subjected to particularly intense scrutiny and application, and that largely in Paris at the very time that Bichat worked there. Bichat used this theory and its attendant epistemology as a kind of binding agent to link together all the other elements of his work.

It is unlikely that Bichat ever went to Auteuil. In fact, he probably had little if anything to do personally with many of the politically-minded *philosophes* whose ideas had been shaped in Madame Helvétius's salon. That made little difference, however, for Condillac's philosophy had become the common property of Parisian scientists and intellectuals. The Société Médicale d'Émulation must have been a particularly important centre for the diffusion of sensationalist ideas. Apart from that, however, all the concepts we have so far examined were widely discussed at the Hôtel-Dieu, as much a part of the institution as the vermin that crept along its walls. There could have been few more intellectually stimulating medical establishments anywhere just at the turn of the century.

To dissect *La vie et la mort* into the elements that compose it is to be brought face to face with notions clearly developed by persons whose work we have been examining. Bichat admitted as much, pointing out that all those persons who had read Aristotle, Buffon, Morgagni, Haller, Bordeu, and others of similar persuasion would see sources of his ideas there. He saw himself, much as we do, as working in a tradition stretching all the way back to Aristotle. It was not his intention to deceive his readers, he wrote (probably quite truthfully), for "those authors are so well known that I thought it useless to note the critical citations exactly."[1] In any case, it is worth remembering that eighteenth-century authors did not tend to be scrupulous about sifting original ideas from borrowed ones. *La vie et la mort*, in spite of its borrowed elements, is important because of what Bichat did with his predecessors' notions. Analytic methodology notwithstanding, the complete work is greater than the sum of its parts. The success of Bichat's published writings and the considerable reputation he achieved in a short time attest to the fact that his synthesis and application of physiological theory were unique. It is significant that when François Magendie and Claude Bernard were striving to create a deterministic science of physiology, they found it necessary to marshal many of their arguments specifically against those of Bichat, who had denied the possibility of what they were attempting.

Bichat opened *La vie et la mort* with the simple assertion: "Life consists in the sum of the functions which resist death."[2] The ephemeral mystery of life and of our consciousness has captivated human curiosity since our mind emerged from a kind of animal semiconsciousness of itself. Bichat dealt with a timeless mystery a bit prosaically, perhaps, by contending that to be alive means fundamentally to be an organized material unit battling subversive forces. With good reason, commentators have generally dismissed the definition as a tautology that enlightens little. It has much in common with Stahl's definition of life as "the conversation of an eminently corruptible body, the faculty or force with whose aid the body is sheltered from the act of corruption."[3] Both men perceived the living body as a kind of organic island, besieged without reprieve by ravaging forces that would dissolve its organization and integrity. According to Bichat, "under such circumstances [living bodies] could not long subsist were they not possessed in themselves of a permanent principle of reaction. This principle is that of life; unknown in its nature, it can only be appreciated by its phenomena." This elaboration does little to dispel the fog, serving no purpose for one desiring something tangible and precise. In fact, it can be quite accurately paraphrased to say that living bodies are alive because they possess the principle of life. In the meantime, in addition to combating the external environment, the life principle directs growth and development. In the child, Bichat wrote, the reactive capacity is superior to the action imposed from without, and there is growth; in the adult, there is a balance between action and reaction; in the old man, life languishes until the reaction overtakes and destroys it.[4]

[1] Xavier Bichat, 'Préface', *Recherches physiologiques sur la vie et la mort*, Paris, Brosson, Gabon, 1800, pp. i–iv. Hereinafter cited as *La vie et la mort*.

[2] Ibid., p. i.

[3] Georg Ernst Stahl, 'Vrai théorie médicale', in *Oeuvres médico-philosophiques et pratiques*, 6 vols., ed. by Theodore Blondin, Paris, Baillière, 1860, vol. 3, p. 43.

[4] Bichat, *La vie et la mort*, p. i.

Every child quickly learns that there are endless varieties of life in the world. He perceives without much apparent difficulty that a plant, a fish, and his pet cat are all alive, as he himself is. This is acquired as an intuition long before he can find words to explain it. Indeed, to describe the thread of reality common to life's myriad species, which ties them together as participants in the experience of life, is very difficult. We all acquire at least some sense of a chain of being that runs through nature, uniting the simpler forms of life by imperceptibly gradated degrees to the more complex ones. To define and to understand this complexity was one of Bichat's principal undertakings in *La vie et la mort*. It was above all the notion of two distinct lives that allowed him to make sense of living diversity. Accordingly, as we have already had occasion to see, he based much else upon it.

What Bichat called the organic life includes all those functions that are normally internal, passive, and unperceived. It corresponds to the plant life of the ancient Greek triumvirate of plant, animal, and rational souls. Its realm is the same one Helmont had assigned to the great integrative *archeus*. When Stahl placed all animal activity under the control of a single rational soul, he actually created conceptual difficulties for subsequent theorists. By not distinguishing between conscious and unconscious functions, animism made the chain of being more difficult to comprehend. Thus it was that Sauvages was led to subdivide the soul into faculties, a device that effectively permitted him to revert to a traditional tripartite division. He believed that the functions Bichat later termed organic were partially, but not exclusively, mechanical. Barthez assigned them to a vital principle. With the exception of Stahl, all believed that there are living functions that exist apart from the conscious psychical life. Above all, they all assumed that the willed, rational and conscious activity is linked to a soul, which is, at the same time, a spiritual entity. Like Bichat, then, all assumed that life has levels into which it must be subdivided to be understood.

Bichat nevertheless considered his particular animal-organic classification to be unique, unrelated to his predecessors' classifications in several important ways. Though Grimaud had stated it clearly,Bichat alone took credit for perceiving its implications. "In reflecting upon the [animal-organic] distinction", he remarked, "I soon perceived that it was not only one of those vast and comprehensive views, one of those great and luminous conceptions that frequently occur to the man of genius who studies physiology; but that it might be made the basis of a methodological classification."[5] In his introductory remarks to the *Anatomie générale*, he contended that Grimaud's particular approach, albeit obviously helpful, lacked precision: that is to say, Grimaud acknowledged only the sensations and motions among the external functions. He did not hold the brain to be their centre and had neglected the voice. Internal functions had been only partially examined, and no consideration had been given to the special status of the organs of generation. In fact, Grimaud's division was merely a minor point in the larger context of his study of nutrition. What was merely a classificatory convenience for him became a key concept for Bichat, central to his very conception of vital function and vital law.[6]

[5] Xavier Bichat, *Anatomie générale appliquée à la physiologie et à la médecine*, 4 vols., Paris, Brosson, Gabon, 1801, vol. 1, pp. xcix–cxii. Hereinafter cited as *Anatomie générale*.

[6] The naturalist Georges Louis Leclerc de Buffon had also written of two lives and of the life proper to

The organic life is the essential one and therefore universally distributed throughout all living things. It was this fundamental quality that Bichat had in mind when he described the plant as "only the sketch or rather the ground work for the animal . . . for the formation of the latter, it has been necessary to clothe the former with an apparatus of external organs by which it must be connected with external objects." When animal life is fused on to it, our organism "senses, it perceives, it reflects on its sensations, it moves voluntarily in response to their influences and it communicates its desires, fears, pleasures and pains."[7] In an 1822 edition of *La vie et la mort*, Magendie commented aptly but tersely that this was a "more brilliant than profound view of the subject".[8]

Bichat found that in the animal, one can distinguish between the organs of the two lives in a number of ways. The parts belonging to the animal life are symmetrical, whereas those of the organic one are irregular in form; animal life is harmonious or regular in its activity, while organic life is not; animal life operates only intermittently, but organic life must be constantly active; animal life is modified by habit, whereas organic life is unaffected by it; rationality belongs to the animal life, while passions are connected to the organic life; animal activity commences at birth, and organic activity is present from the moment of conception; animal life alone can be educated. Because animal life cannot exist without organic life, the former leaves the body first at death.

He had at least the outlines of this distinction as early as 1798, for that year he published in the *Mémoires de la Société Médicale d'Émulation* a paper concerning the differences in form between the organs of two lives. Whereas a tree has only random branches on an irregular trunk, animal limbs are always symmetrically arranged on an irregularly shaped body. A single organ of the animal life is always placed along or in relation to a median line that divides the body into two halves. Thus it is that two similar eyes receive light impressions, and the tongue is split by a median line. The nerves that transmit sensory impressions from the eyes, ears, and so on to the brain are symmetrical, as are those which go from the brain to the larynx and locomotive organs. The brain itself exists in two apparently identical parts. On the other hand, the digestive organs, liver, spleen, heart, aorta, vena cava, and other vessels are not absolutely regular.

But all the evidence did not quite fit, and Bichat had to do some verbal juggling to salvage what he was convinced was a physiological principle. The apparent regularity of the glands and kidneys, he warned, is merely illusory. One human lung, for example, has two lobes, and the other three. Furthermore, the nerves and vessels supplying the two sides vary considerably in shape and direction. The pancreas, liver, and salivary glands are not quite on the median line, and so on. The reproductive system participates in both lives and is, as one might expect, symmetrical.

---

each part of the body. This idea belongs to the concept of the chain of being, a belief in the continuity of life, which the biologist attempts to classify into order, genus, and species. The connexion between Buffon and Bichat, was first drawn in A. Arène, 'Essai sur la philosophie de Xavier Bichat', *Arch. Anth. Crim.*, 1911, **26**: 32–35.

[7] Bichat, *La vie et la mort*, pp. 1–9.

[8] François Magendie, note to Bichat, *La vie et la mort*, 4th ed., augmented with notes by F. Magendie, Paris, Bechet, 1822, pp. 6–7.

In other instances, Bichat's examples in support of his viewpoint are rather intriguing and in some sense more persuasive. He interpreted the evidence surrounding symmetry, for example, in such a way as to determine that it increases in proportion to a species' location on the chain of being. Man, nature's allegedly most perfect animal, is also its most symmetrical one. The lowly oyster, on the other hand, is irregular and confined to the shell that covers it, preventing it from having any relation with external bodies. Caterpillars and butterflies are symmetrical. When they enter the intermediate or cocoon stage of their existence, however, they collect themselves into irregularly shaped shells and live completely internalized lives.[9]

The somewhat cynical reader may not be surprised at Bichat's claim that the symmetry principle was original to him. I believe, however, that the catalyst for the idea was Bordeu's *tissu muqueux*, also known in the eighteenth century as cellular tissue. We generally know it today as connective tissue. In the work, Bordeu described a *raphé générale*, a kind of constriction that divides the body into right and left sides. The diaphragm is divided into muscles that are directed toward the spine and sternum; there is a mediastinum; a division is evident in the trachea, thyroid gland, thyroid cartilage, cricoid, epiglottis, and nostrils; the oesophagus and pharynx are separated posteriorly by a line or crimping together of fibres; the tongue has a median line; the vertebrae and even the intestines are divided into demicanals; the brain, pineal gland, liver, and pancreas are located in the centre of the body and divided into two parts; the mucous tissue itself is constricted at its centre and so divided as to form a pocket on either side of the body.[10] There is good reason to suppose that whenever Bichat referred critically to Bordeu's work, he was in fact borrowing and developing it somehow. Thus it was that in 1798, he accused Bordeu of making a forced and unnatural application of a worthy principle. If one merely glances at the list of organs offered above by Bordeu, it is clear that he did not limit symmetry to the organs of what Bichat called the animal life. And Bichat had the advantage of the last word.

The symmetry of the organs of the animal life is paralleled, according to Bichat, by harmony of function. Buffon had observed that there is a harmony of action of the eyes and ears. There is a discordance in the voice, he said, if there is a discordance in the two halves of the larynx. From similar considerations, Bichat derived a second "principle of life", which stated that "harmony is the character of external functions while discordance is the attribute of organic functions". We sense confusedly, he wrote, if one eye or ear or nostril is stronger than the other or if one part of the body is affected by paralysis or spasm. A lack of harmony in the hemispheres of the brain will cause the soul to perceive confusedly. He observed that compression of one of the hemispheres by pus or blood is sometimes known to produce confusion in persons and in animals.[11] Whereas the fact that one is always left- or right-handed seems to contradict the notion of harmony, Bichat put the phenomenon down to habit. Harmony is not important, however, for the organic life. It makes no difference, for example, if

[9] Xavier Bichat, 'Organes à forme symmétrique', *Mémoires de la Société Médicale d'Emulation* 1798, **2**: 477–487; and *La vie et la mort*, pp. 10–19.

[10] Théophile de Bordeu, 'Recherches sur le tissue muqueux ou l'organe cellulaire', in *Oeuvres complètes*, 2 vols., Paris, Caille et Ravier, 1818, vol. 2, pp. 753–755; also briefly discussed in 'Recherches sur les maladies chroniques', in ibid., vol. 2, pp. 801–802.

[11] Bichat, *La vie et la mort*, pp. 20–36.

one kidney is more active than the other or if one salivary gland secretes more saliva than the other. Furthermore, the activity of the glandular system, the circulation, or the respiration is constantly varying as much as twofold or threefold under the influence of various causes.[12]

As inharmoniously as the organic functions may occur, however, they must never be interrupted. It is clear that respiration, circulation, exhalation, absorption, nutrition, assimilation, and decomposition must be continuous. The organs of the animal life, on the other hand, are subject to fatigue and must periodically relax. Rest may be the repose of a single fatigued organ or, when the brain is resting, a generalized sleep.[13] As we have observed, this notion came directly from Grimaud and indirectly from Barthez, who had stated that only organs "necessary to life" enjoy continual excitation by the vital principle.[14]

A particularly interesting part of Bichat's examination of the two lives had to do with the process of learning. Only the animal life can be taught, he wrote. His assumptions were largely sensationalist ones. Like many other men at the time, he believed that habit is of primary importance in permitting one to learn something of some object. An initial impression, Bichat wrote, is merely agreeable or disagreeable. It is confused and inexact until such time as we begin to decompose it into its parts. To illustrate this point, he used the example of a field of flowers, much as Condillac had offered the view of a countryside. Bichat considered this analysis of the components of some larger whole to be an exercise of judgement.

Unfortunately, he went on, there is an inverse relationship between appreciating something and understanding it. "The more we consider an object, the less sensitive we are to its agreeable or disagreeable qualities, but at the same time, the better we may judge its attributes." We become accustomed not only to a beautiful view but also to an irritating foreign body in contact with a mucous membrane.[15] Though we are aware of a sudden passage from hot to cold surroundings or the other way around, we soon cease to notice the new temperature. Neither the perfumer nor the cook is sensible of the odours surrounding him. This dulling of sensations by habit is not a function of the sensory organs, but rather of the mind itself, which compares each sensation with preceding ones. "The greater the difference between the actual and the past impression, the livelier will be the sentiment. The sensations which affect us the most are those which we have never before experienced." Thus pain and pleasure tend to their own annihilation. The poets' sentiments, it appears, must retreat before physiological processes, for there can be no eternal sorrows.[16]

These words make it clear that Bichat knew Cabanis' ideas well and found them to be persuasive. Some of Bichat's handwritten lecture notes are preserved at the Library

[12] Ibid., pp. 36–38.

[13] Ibid., pp. 39–46.

[14] Paul Joseph Barthez, *Nouveaux éléments de la science de l'homme*, Montpellier, J. Martel ainé, 1778, pp. 235–244.

[15] Because mucous membranes line organs in contact with the exterior, such as the bladder and the gut, Bichat taught that it was possible that they shared some of the characteristics of the animal life. Hence these organs respond to habit.

[16] Bichat, *La vie et la mort*, pp. 47–56; compare Étienne Bonnot de Condillac, 'Essai sur l'origine des connaissances humaines', in *Oeuvres complètes*, 23 vols., Paris, Ch. Houel, 1798, vol. 1, pp. 157–172.

of the Paris Medical Faculty, among them twenty-nine pages entitled 'Discours sur l'étude de la physiologie'. They tell us that Bichat taught his students that "all the sciences are divided into the moral and physical sciences", an idea clearly taken from Cabanis. After a note concerning life's division into two categories, he instructed himself, "Speak here of Cabanis", and observed, "Since Locke and Condillac found the source of our ideas in the senses, it is essential to know these senses".[17] The connexion between some of Bichat's ideas and those of Cabanis or the other ideologues is made abundantly clear in *La vie et la mort* also.

The question of the role of habit in the intellectual operations interested both Destutt de Tracy and Cabanis, both of whom read papers on the subject before the second class of the Institute in 1798. In the *Éléments d'idéologie*, which appeared the same year as *La vie et la mort*, Destutt de Tracy argued that habit affects our sensations, motions, memories, desires, and judgements. As we have seen, Cabanis treated the subject again in *Rapports du physique et morale de l'homme*.[18] In fact, he accused Bichat of plagiarism in the preface to the book. While discussing the general topic of the science of man, he appended a footnote stating that he had just heard of Bichat's death, and sharply accusing persons who, without scruple, get hold of others' ideas but neglect to indicate their sources.[19] His sense of injustice probably stemmed largely from this element of Bichat's work.

In 1803, the Institute awarded a prize to one Maine de Biran for a paper on 'L'influence de l'habitude sur la faculté de penser'. Maine de Biran was sharply criticized subsequently for not having named Bichat in the work. He defended himself by saying that when he began his work in 1799, *La vie et la mort* had not yet appeared. On seeing Bichat's book later, he was pleased to see the basis of an idea that he had thought to be exclusively his own.[20] Such confusion about priorities arose largely because certain ideas floating about medical circles were there for anyone who cared to develop them. Destutt de Tracy, Cabanis, Bichat, and Maine de Biran were all in Paris about 1800, and they all had occasion to hear something about the relationships between sensations, the internal organs, will, habit, and intelligence.

Habit, according to Bichat, belongs only to the animal life, however. Indeed, life would clearly be menaced if visceral organs were thus affected. Its effect on the organic life is minor, therefore, limited to modifying the hungry stomach and the excretory organs. Even in those two cases, however, Bichat claimed that it does so in proportion to the extent to which these particular functions participate in the animal life because of their contact with such foreign substances as food or excrement.[21]

[17] These notes by Bichat were reprinted as a 'Discours sur l'étude de la phisiologie', ed. by A. Arène, *Arch. Anthr. Crim.*, 1911, **26**: 161–172. The original manuscript in Bichat's almost impossible hand is in the Bibliothèque de l'École de la Médecine, Box 46, no. VIII. It is translated by William Randall Albury as the 'Discourse on the study of physiology', *Stud. Hist. Biol.*, 1977, **1**: 97–105.

[18] Pierre Jean Georges Cabanis, *Rapports du physique et du morale de l'homme*, 4th ed., 2 vols., Paris, Bechet, 1924, vol. 1, p. 187.

[19] Ibid., p. xiv.

[20] That letter was discovered in this century and reproduced by V. A. Bertrand in 'Xavier Bichat et Maine de Biran', *Arch. Anthr. Crim.*, 1911, **26**: 434–443. The dispute was also discussed by Arène, op. cit., note 6 above.

[21] Bichat, *La vie et la mort*, pp. 56–57.

Because of his internal organs, Bichat's man or animal was very different from Condillac's statue, which possessed only the organs of what he called the animal life. Condillac had described a passion as a dominant desire deriving from the memory of a pleasant sensation that the statute wished to recreate.[22] Destutt de Tracy and Cabanis had allowed that internal sensations may modify external ones. Bichat, however, set out to demonstrate that the passions – anger, love, joy, sorrow, and so on – originate in the organic life. Without external sensations to put him in contact with the larger life, of course, man would be bereft of all intellectual activity. But without the internal sensations that give rise to the passions, he would presumably lack the qualities that shape the personality and provide life with its colour. The two lives are occasionally at war in the body, with the passions opposing the judgement and intellect of the animal life.

Anger, fear, and joy all variously affect the heart rate and the circulation; respiration is oppressive when one is very sad; certain passions affect the digestive system, causing vomiting or indigestion. That, wrote Bichat, is because these are the sources of the respective passions. Even gestures, he claimed, attest to the connexion between the passions and the organic life. To indicate an expression of joy, love, hatred, or sadness, one gestures toward the heart, the stomach, or the intestines. Popular expressions to the effect that fury circulates in the veins and stirs up the bile, that anger makes the heart leap, and that jealousy distils its passions into the heart, wrote Bichat, owe as much to physiology as to poetry.

It follows, naturally enough, that organic dispositions contribute to the personality. A person with a strong pulmonary apparatus and energetic circulation possesses what has long been labelled a sanguine temperament, which disposes him to anger and to courage. Envy and hate are more pronounced in someone in whom the bilious system dominates. The opposite of the impetuous, sanguine man is the inactive and dull one, in whom the lymphatics are greatly developed. The imagery of the temperaments is, of course, at least as old as Aristotle. If one accepts this viewpoint, it necessarily follows that man is unavoidably assisted or incapacitated by his unique and largely unalterable physiological make-up.

Human passions alter over a lifetime. Childhood, Bichat wrote, is an age of timidity, because organization is feeble. In youth, the pulmonary and vascular systems are more developed, presumably somehow accounting for the qualities of one's personality in those years. Maturity is the age of virility, ambition, envy, and intrigue – all somehow having to do with the alleged fact that the activity of the liver and gastric apparatus is pronounced. Although plants possess an organic life, they lack passions, because they have neither sensory apparatus nor appropriate viscera.

Even though they originate in the organic life, the passions frequently affect the animal life. An angry man's muscular force is doubled or even trebled because of an increase in the power of his heart, and hence in the quantity of circulating blood. Indeed, passions produce a thousand involuntary nuances of force, even in voluntary muscles. These effects are the result of sympathies, imperceptible connexions that exist between two remote and apparently unrelated parts of the body. Bichat probably borrowed this notion and even the term from Barthez. Therefore, while it would seem

[22] Condillac, 'Traité des sensations', in *Oeuvres complètes*, op. cit., note 16 above, vol. 1, pp. 90–95.

that the brain ought not to be affected by the liver, stomach, or intestines, there are numerous sympathies between these organs. They must exist also between the viscera and the voluntary muscles themselves.

The intellect and the passions, therefore, coexist in a kind of balance in the body. For example, Bichat wrote, a man who gets bad news when he is before a crowd constrains his normal responses. "It is the brain whose action has surmounted that of the stomach, the liver and so on; it is the animal life which has reclaimed its empire." The happiest man is one in whom the two lives are in equilibrium, so that the cerebral and internal centres exercise an equal control.[23] This discussion brings to mind the fourth book of Jonathan Swift's *Gulliver's travels*, in which the hero visits the land of the Houyhnhnms. These animals, much like horses in appearance, lead superbly rational and ordered lives. They achieve their unparalleled comfort and sobriety, however, because they are never deflected by passions or emotions. Their opposites are the vulgar, unpredictable, violent, and emotional Yahoos, who bear an unmistakable physical resemblance to humans. Bichat probably knew nothing of Swift, but in effect he provided a kind of physiological rationale for his fanciful and satirical venture into the realm of the psyche.

Bichat found that he could distinguish between the two lives on the basis of their origin and development as well. The organic life, he wrote, is active from the moment of conception, but animal life does not truly begin until the senses, the brain, the organs of locomotion, and the voice begin to be exercised after birth. Condillac's statue began to develop its mental faculties only when it had two odours, two temperatures, or any other two sensations to compare. In the womb's constant milieu, the foetus has no consciousness of the medium that nourishes it or the heat that penetrates it. Its motions are unconscious and unwilled, deriving from sympathies between internal organs. The very existence of the foetus, Bichat wrote, "is that of a vegetable and its destruction can only be said to be that of a living being and not of an animated one." It followed that wherever there is any question of saving a mother's life or that of her unborn child, the decision is properly made on the mother's behalf.

Bichat must have had Condillac's work near at hand when he wrote the following lines concerning the newborn:

> The sensations are at first confused; they transmit only general images; the eye has only the sensation of light; the ear has only that of general sound; the nose only that of smell. As yet, there is nothing distinct in the general affections of the senses; but from habit the strength of the first impression is lessened and particular sensations begin to take place.

So the animal goes on to develop its mental faculties. "The powers of perception, memory and imagination all of which are preceded and occasioned by the sensations increase and extend in proportion as by repeated excitement they are exercised." At first, the cries of young animals are only confused and unformed sounds with no particular character. Children affect the characteristic sounds of their species only after long effort. In a newborn animal, the limbs are constantly in motion, as it attempts to experience more sensations. Only in time does it learn co-ordination of its muscles.[24] In the 1822 edition of *La vie et la mort*, Magendie sensibly pointed out, for

[23] Bichat, *La vie et la mort*, pp. 58–91.
[24] Ibid., pp. 134–151.

example, that a ewe can distinguish the voice of her lamb in the flock from the very beginning. As any farmer knows well, there is no question of an animal's having to learn to make and to distinguish the sounds of its species. Nor do the motions of the locomotive organs always require any more education than does the heart. A partridge, for example, emerges running from its shell.[25] Obviously, there are countless similar examples to be had, if one but thinks about the problems for a while.

Bichat divided the education of the external life into three categories: the senses, the brain, and the body. The first division includes the visual arts, music, cooking, perfumery, and so on; the second has to do with poetry and the sciences; the third concerns riding, dancing, and the mechanical arts. They exist in a reciprocal relationship, he wrote, so that while the philosopher is often physically awkward, the dancer is usually not too bright. This provided him with yet another principle, the "fundamental law of the distribution of vital powers". He believed that a "determinate sum of forces or powers have been spread throughout life; this sum must always remain the same whether its distribution is equal or not; the inactivity of one organ necessarily supposes activity in another." To apply oneself to several studies at once is probably to succeed at none of them. Another consequence of the principle is that while hearing and touch are strong in a blind man, the deaf and dumb man has more accurate eyesight. An eagle has piercing sight but an obscure sense of smell, while the opposite is true of the dog. When vital powers are centred in one life, they are relatively inactive in another. When digestion is going on, for example, the powers of life are centred in that system, and the animal becomes sleepy. Bichat clearly would have considered football scholarships inappropriate for a university.

As a consequence of this same law of the distribution of living forces, Bichat said, various organs are perfected at different stages of life. In infancy, the senses are particularly educated, and the nervous system is proportionately greater in relation to the muscular system than later. In youth, memory and imagination become more active, while in adulthood, the faculty of judgement is developed. In view of this, we would do well to apply children to dancing and music, young people to the fine arts, and the adult to logic and mathematics.

Organic life, ineducable and unaffected by habit, is present from the moment of conception. The heart is the first organ to be formed and to begin acting in the foetus. At the beginning of life, the vital powers are concentrated on growth and nutrition. Exhalation, respiration, and digestion begin only after birth.[26]

Bichat did not discuss the possible social consequences of his physiological theory, though other persons would do so. I commented earlier that it seems to follow from the principles of Locke and Condillac that all human beings possess an equal potential at birth. The infinite variability of our species presumably springs from the complex effects of the environment. This was also the viewpoint of such ideologues as Helvétius, Turgot, and Condorcet. Frank Manuel has shown that it was really only during the French Revolution that Frenchmen began to abandon the seductive notion of innate human equality. Cabanis presented papers before the Class of Moral and Political Sciences of the Institute, arguing that man's physical and moral being is

[25] Magendie, op. cit., note 8 above, pp. 204–205.
[26] Bichat, *La vie et la mort*, pp. 147–177.

widely affected by age, sex, climate, and many other such variables.[27] Now Bichat would have it that the human temperament and moral character are consequences not only of the environment but also of one's spleen, heart, lungs, and gastrointestinal tract. At best, one can hope to modify one's character by training the will and judgement to moderate impulses coming from the passions of the uneducated internal life.

One is also assisted or fettered, according to Bichat, by the limitations of the animal life. We observed earlier that to receive a sound education, one must concentrate upon the activities of the senses, the brain, or the muscles. One cannot be a good musician, philosopher, and athlete. The division of the animal life into three parts was developed in a most interesting way during the Restoration by the Comte de Saint-Simon, whose political and social philosophy was rooted in a notion of human inequality. He first discovered Bichat's work about 1822 and used it to postulate a hypothetical society in which people were to be channelled according to their natural abilities. "Brain men" would become scientists; "sensory men" would be poets, religious leaders, and ethical teachers; "motor men", the intellectually mediocre majority, would be either industrial workers or administrators.[28]

Finally, Bichat showed that the two lives remain distinct from each other even at death. An ageing person dies gradually. Hearing, sight, taste, and touch fade slowly but perceptibly. Imagination, perception, and memory slow. The muscles of locomotion and the voice falter as the brain becomes less and less active. The imagination weakens, so that often an old man can recall events long past while being unaware of the present. While this goes on, the organic life continues unabated. We fear death, Bichat observed, because we fear extinction of our consciousness. This fading of the animal life, therefore, is really an advantage, for one gradually approaches the vegetable state, thereby removing fear of extinction.[29]

The animal-organic division was a device that frequently permitted Bichat to reconcile the apparently conflicting notions and conclusions of such physicians as Haller and the Montpellier vitalists. It provided him with a kind of conceptual framework around which he proceeded to build his arguments on behalf of vital theory and tissue theory. In all the work he did subsequently in physiology and anatomy, he never lost sight of what he saw as a critical natural division.

One of the most important and instructive features of Bichat's work for the medical historian is his theoretical statement on behalf of the rejection of mechanism and the

[27] Frank E. Manuel, 'From equality to organicism', *J. Hist. Ideas*, 1956, **17**: 54–69. See also Manuel's *The prophets of Paris*, Cambridge, Mass., Harvard University Press 1962, pp. 105–148.

[28] Manuel (1956), op. cit., note 27 above. This provoked a novel approach to Bichat's physiology on the part of John V. Pickstone, 'Bureaucracy, liberalism, and the body in post-revolutionary France: Bichat's physiology and the Paris School of Medicine', *Hist. Sci.*, 1981, **19**: 115–142. Striving mightily for a social interpretation of the apparently apolitical Bichat's work, Pickstone argues that Bichat, like Saint-Simon, saw the bodily parts through an imagery of state organization so that his physiology represents a kind of "bureaucratic corporitism" in which the anatomical elements were also functional elements. On pp. 133–134, he gives an organizational chart comparing Bichat's physiology and a model of social organization characteristic of his time, claiming that a "social physiology was a plausible representation of official France during the Directory". He even compares Bichat's tripod of life to the interactions of the Directors, each with their separate executive functions. The "linguistic habits" to which he attributes this convergence of ideas notwithstanding, I find the argument facile, albeit ingenious.

[29] Bichat, *La vie et la mort*, pp. 178–189.

support of vital theory. By this time, there were few defenders of traditional mechanism. The eighteenth-century medical world belonged to physicians who did not doubt that living functions are far more mysterious, complex, and elusive than mere clockwork. Bichat offered his readers a concise and articulate theoretical statement in support of a widely assumed position. It can be summed up by saying that the physical sciences are concerned with regular and hence predictable phenomena, whereas the life sciences deal with infinitely variable and constantly altering events and situations. His arguments, in their own terms, were both sensible and compelling, and most contemporary medical men assented to them. Indeed, they were taught in the medical schools well into the nineteenth century.

Bichat wrote as follows concerning the question of predictability and variability in the natural world:

> [Vital powers] vary incessantly in their intensity, energy, development; they are constantly passing from the lowest degree of prostration to the highest pitch of exaltation and they assume a thousand modifications under the influence of the most trifling causes. For the animal is influenced by everything around him; he wakes, he sleeps, rests or exercises himself, digests or is hungry, is subject to his own passions, and to the actions of foreign bodies. On the contrary, the physical laws are invariable, always the same and the source of a series of phenomena which are always the same. Attraction is a physical power; it is always in proportion to the mass of brute matter in which it is observed; sensibility is a vital power, but in the same mass of matter, in the same organic part, its quantity is perpetually changing.

With mathematics, he went on, one can calculate the return of a comet or the resistance of fluid passing through an inert canal. But to calculate the force of a muscle with Borelli, the velocity of blood flow with Keill, or the quantity of air in the lungs with Lavoisier is to build a solid house on shifting sand. One never knows vital fluids after a single analysis, because urine, bile, and saliva vary throughout the day, during disease, and with age. "The instability of the vital powers is the quicksand upon which the calculation of all the physicians of the last century has sunk." The major problem with physiology, Bichat contended, is that it was developed only after physics. Had it been cultivated before, men might have made applications from the former to the latter, rather than the other way around. In that case, rivers might have been seen to flow from the tonic action of their banks, crystals to unite from the excitement they exercise upon their reciprocal sensibilities, and planets to move because they mutually irritate one another at great distances. Today, we would consider the application of such language to the phenomena of physics a reversion to a more primitive, even animistic conception of nature. Bichat, however, was not looking backward toward a simpler or more superstitious age, but rather forward to one in which medical studies would attain a new level of theoretical coherence and methodological sophistication.

Most vitalists were prepared to concede some accessory status to physics and chemistry in the study of life. Bichat, however, emphatically denied even that, claiming that those sciences are wholly alien to it:

> There are two classes of beings in nature, two classes of property and two classes of science. Beings are organic and inorganic, properties vital and non-vital, sciences physiological or physical. The vital properties are sensibility and contractility. The non-vital ones are gravity, affinity and elasticity.[30]

In his handwritten lecture notes, he said of Borelli's application of mechanics to the

[30] Bichat, *Anatomie générale*, pp. 1–2.

study of muscular forces that "The calculations are correct, but the principle from which he begins is false because in each instance, the forces in muscles vary." It is a mistake to consider, as did Boerhaave, that arteries, veins, and other canals in the body act as inert tubes that behave in a precise and predictable manner. Although the physiologist must know something of the laws of optics and acoustics, he must never assume that vital laws and living activity are subject to calculation. Mathematics, therefore, is of no use to physiology. "Insist here", he reminded himself, "on the uselessness of mathematics."[31]

Bichat imitated Barthez' language on many occasions when he dealt with the subjects of matter, natural law, and the role of the scientist. His views on matter and its properties were unquestionably shaped by what Barthez had written on the subject. Using words almost identical to those of the Montpellier vitalist, Bichat wrote, "Let us pay homage to the immortal Newton; he was the first to find the Creator's secret, namely that of uniting a simplicity of causes to a multiplicity of effects." As we observed earlier, Barthez had set himself apart both from physicians, who enumerated large numbers of causes for physiological phenomena, and from the animists and mechanists, who tried to assign all living activity to a single cause. Somewhat perversely, Bichat turned Barthez' own words about the "simplicity of causes" and "multiplicity of effects" against him, accusing him of the same fault as that of the dualists.[32] He described the vital principle as an empty word, "an assumption as void of truth as to suppose one sole acting principle governing all the phenomena of physics. Amongst the latter, some are derived from gravity, others from elasticity, and still others from affinity. So in the living economy, some result from sensibility, others from contractility and so on." Unable to conceive of these properties as components of matter, Barthez had placed them in the vital principle. With some justification, Bichat merely dismissed as a chimera this notion of a substance that is neither soul nor matter.

However we might be inclined to interpret it, Bichat sincerely believed that his own particular analysis of the body as governed by vital forces was superior to those of all his predecessors. Indeed, he was largely correct when he implied that they had frequently confused words with substantially demonstrated causes. The trouble with the *archeus*, the soul, and the vital principle, he wrote, is that each one in turn was taken to be the common basis of every physiological explanation, a kind of first cause. Certainly Stahl and Barthez, at least, would have denied searching for or naming first causes. Indeed, the charge was a particularly nasty one to aim at Barthez, who saw himself as a staunch Newtonian. And as we observed, neither Helmont nor Barthez

[31] Bichat, op. cit., note 17 above. A somewhat differently slanted interpretation of Bichat's attitude to the mechanical sciences is found in Geoffrey Sutton, 'The physical and chemical path to vitalism: Xavier Bichat's *Physiological researches on life and death*', *Bull. Hist. Med.*, 1984, **58**: 53–71. Largely examining Bichat's experiments on the death of the heart, lungs, and brain, Sutton emphasizes Bichat's preoccupation with the *mechanical* action of the heart, which sustains the life of the tissues. The centre of the organic or fundamental life and especially of insensible organic contractility, the heart is the centre of mechanical agitation. The activity of the brain, the centre of the animal life, however, is different in its nature. To demonstrate that, Sutton contends, Bichat was working with the new techniques of galvanism. Interesting and perceptive though his emphasis is, however, it does not detract from the insistence of Bichat on the two sciences.

[32] These arguments are discussed in Réjane Bernier, 'La notion du principe vitale de Barthez', *Archs. Phil.*, 1975, **35**: 423–441; and in Elizabeth Haigh, 'The vital principle of Paul Joseph Barthez: the clash between monism and dualism', *Med. Hist.*, 1977, **21**: 1–14.

really treated the *archeus* or vital principle respectively as sufficient explanation for all living phenomena, but only for the ones Bichat assigned to the organic life. Bichat's objections had really to do largely with their use of words. "*Archeus*" or "vital principle" sound very much as though they represent causes. Bichat found such theoretical props to be unacceptable, for, tending to seduce people into substituting names for causes, they allow them to believe that they have accounted for those causes, whose nature nevertheless remains unknown. Although one has to suppose that general causes exist, he wrote, the scientist's eye should be directed to observing their results.[33] He appears not to have noticed that this was the very thing Barthez had proposed to do.

According to Bichat, physical properties are permanently attached to matter, whereas vital ones are only temporarily imposed upon it. Chaos, he wrote, would be a state in which matter is devoid of all properties. At the creation, God endowed the universe with gravity, elasticity, affinity, and so on. He animated a portion of it by adding the properties of sensibility and contractility to the physical ones. Life, then, is added on to inert matter. This occurs when an object is brought into contact with something alive, thus undergoing a kind of injection of vitality. "In passing from time to time through the living bodies, matter is penetrated at different intervals with the vital properties which are found in combination with the physical ones." While they are present, the vital properties dominate, so that "fettered by vital properties, the physical ones are continually restrained in the phenomena which they tend to produce." Bichat and Barthez here were divided largely on the question of where the sensible and contractile forces are located, the former assigning them to the tissues and the latter to his hypothetical vital principle. In spite of relying so heavily on Barthez' arguments about material properties, Bichat rarely mentioned his work except to dismiss it. On this occasion, he offered merely a patronizing nod in the direction of Montpellier. "Our art is deeply indebted to several physicians of Montpellier for having driven the Boerhaavian theories from the schools and for having embraced the opposite ones of Stahl", he wrote, "but in departing from the false track which they were pursuing, they have chosen in its place such a wild and tortuous path that I doubt much if they will ever find an outlet."[34]

The vital properties, five in all according to Bichat, derive from the forces of sensibility and irritability, which God conferred upon living matter. These properties incorporate the data and speculations accumulated by Haller and by Bordeu. By imposing his versatile notion of the two lives upon their theories and insights, Bichat showed that their differences of opinion concerning the property of sensibility were not substantial. Both men, he said, were correct in their observations and in their basic conclusions. Their disagreement simply stemmed from the fact that they were talking about the properties of two different lives. Bodily parts do possess unequal amounts of sensibility and contractility, as Haller showed. They also owe their myriad automatic functions to another sort of specific sensibility and contractility, as Bordeu pointed out. Bichat's system even managed to resolve the conflict between Haller and Whytt, whose observations often contradicted each other. The point, as he saw it, is "that if

[33] Bichat, *La vie et la mort*, pp. 92–98.
[34] Bichat, *Anatomie générale*, pp. v–ix.

these divisions had been clear and precise, if the words sensibility, irritability, tonicity, etc., had all been used in the same sense by all, we would find in the writings of Haller, of Lecat, of Whytt, of Haen, of all the Montpellier physicians, etc., few of those disputes which were of no importance to science and tiresome for those who study them."[35]

The point, very simply, is that perception and motion exist in each of the two lives. Their character in the animal life differs in many respects, however, from that in the organic life. The following chart outlining the properties of living bodies is taken from *La vie et la mort*.[36]

Classes	*Genus*	*Species*	*Varieties*
Vital	1. Sensibility	1. Animal	
		2. Organic	
	2. Contractility	1. Animal	
		2. Organic	1. Sensible
			2. Insensible
Of tissue	1. Extensibility		
	2. Contractility		

The sensibility of the organic life consists in the faculty of receiving an impression in a particular bodily part. It was this vital property that Bordeu treated when he examined glandular functions. It is equivalent to Glisson's *perceptio naturalis*. In the animal life, however, the impression is referred to the brain, where it becomes conscious. When an animal responds to the abuse of some part of its body with cries of pain, it is demonstrating animal sensibility, the property Haller described in his paper 'On the sensibility of the parts'. Bichat distinguishes the two sensibilities as follows:

> The stomach is sensible to the presence of food, the heart to the stimulus of blood, the excretory tube to the contact of the fluid which is peculiar to it; but the term of this sensibility is in the organ itself. In the same way do the eyes, the membranes of the nose and mouth, the skin and all the mucous surfaces at their origin receive an impression from the bodies which are in contact with them, but afterwards they transmit such impressions to the brain which is the general centre of the sensibility of these organs.

The greater the amount of animal sensibility it possesses, the more "perfect" the animal, for this vital property is the source of the sensations, the perceptions, and the pleasure and pain that regulate them.

For the same reasons as Whytt, Bichat found Haller's classification of the bodily parts into sensible and insensible categories to be too rigid. For example, Haller believed ligaments to be insensible to painful stimuli. Bichat found that while they do not respond to acids, alkalis, or cutting, they are pained when twisted, torn, or distended. Though an animal is not conscious of the blood in its arteries, it cries when a foreign fluid is injected, and so on. But Bichat's division of sensibility into two categories was also too rigid. To make sense of the evidence, he had to admit that animal and organic sensibility are the same in their essential categories, with animal sensibility being a kind of maximum of its organic counterpart. Inflammation can increase organic sensibility to a painful level. It becomes, thereby, animal sensibility. On the other hand, habit can reduce animal sensibility to an organic or unconscious

[35] Bichat, *La vie et la mort*, p. 133.
[36] Ibid.; see also Bichat, *Anatomie générale*, pp. xi–xiii, lxxii–lxxix.

level. Every organ possesses sensibility, Bichat wrote, but there are a thousand degrees of it between that limited to the organ itself and that perceived by the brain. This admission, it would seem, should have detracted from his commitment to the animal-organic division, but we find no evidence that it did so.

Like Bordeu, Bichat attributed each organ's activity to its own particular sensibility. Because of it, the larynx is closed to everything but air, the intestinal lacteals absorb only chyle, and so on. Alteration of the normal level of sensibility is disease. Serous surfaces, for example, bathe for months in fluid they do not imbibe. If their sensibility should be increased by inflammation, however, a new level comes into equilibrium with that fluid, and absorption takes place, producing oedema.[37]

The two contractilities do not shade into each other, as the sensibilities do. Animal contractility is subject to the will. It is located exclusively in the voluntary muscles and functions in conjunction with the soul, the brain, and the nerves. Organic contractility does not depend on a common centre but exists only in the moving organ itself. Both are intimately linked to their corresponding sensibilities. In the animal life, the brain and nerves transmit the signals between sensibility and contractility. Because the vital properties exist in the same structures in the organic life, such transmission is not necessary.

The vital property of organic contractility provided Bichat with some difficulties. He found that he had to subdivide it in order to have it perform all the functions it ought to. Sensible organic contractility, wrote Bichat, controls organs such as the heart, stomach, intestines, and bladder. An insensible organic contractility governs unperceived functions such as those of excretory ducts, secretion, lymphatics, and other small organs where fluids are disseminated in very small quantities. Indeed, the latter vital property is exercised upon mere molecules of matter. The two subdivisions ought nevertheless to be classified together, he believed, because in spite of external appearance, both types pertain only to the internal life and act independently of the will. Indeed, they are connected by indeterminate gradations. Unlike the case with sensibility, however, organic contractility can never be transformed into its animal counterpart. For example, in no case can intestinal movements ever be made subject to the will.[38] The insensible organic contractility acted much like irritability, which in Bordeu's terms, provokes a gland's sensibility, causing it to secrete its humour. It was also closely related to the *motus naturalis*, which, according to Glisson, accompanies the *perceptio naturalis* in the unconscious activities of the body. Clearly, animal contractility is the same as the irritability Haller discovered in the voluntary muscles, while the sensible organic contractility performs the same functions as the *vis insita*.

In the *Anatomie générale*, Bichat asked whether sensibility is a necessary component of contractility. The work of Glisson, Bordeu, and other Montpellier physicians treated them as inseparable, whereas Haller had not. His particular system of vital properties, said Bichat, set this dispute at rest, along with many others. As in the case of other arguments surrounding the two properties, his predecessors were all correct. "First, in the animal life", Bichat wrote, "it is clear that contractility is not a necessary consequence of sensibility; thus external objects often act upon us for a long time and yet

[37] Bichat, *La vie et la mort*, pp. 99–111.
[38] Ibid., pp. 112–121.

the voluntary muscles are unmoved. On the other hand, in organic life, these two properties are never separated."[39]

Bichat described two other properties of living tissue that are not, strictly speaking, vital properties. That is, they do not leave animal tissue at death, but only when it has decomposed. Thus they are, Bichat believed, more like properties of organization. The extensibility and contractility of texture are observed when the skin stretches with tumours, obesity, or pregnancy, when the dura mater and bones of the cranium are extended in hydrocephalus, and when abscesses or haemorrhages occur in the viscera. When such extension ceases after weight loss, childbirth, draining of abscesses, and so on, the contractility immediately sets in. Many persons, he wrote, including Haller and Barthez, had confused it with the phenomena properly belonging to the insensible organic contractility.

The vital properties are not equally distributed throughout the various parts of the body. A voluntary muscle possesses all the varieties of contractility. When it is activated by nerves, it undergoes animal contractility; when it is excited to contract by a chemical or physical agent directly applied to it, it undergoes sensible organic contractility; when fluids enter it to provide nutrition, they provoke a small vibration in each fibre which is the insensible organic contractilty; when the muscle is severed transversely, the parts retract toward their points of insertion because of their contractility of texture. The heart and intestines lack animal contractility. Such organs as tendons, aponeuroses, and bones possess only insensible organic contractility and contractility of texture, and so on. Finally, a physiologist need not have recourse to any forces but those of sensibility, contractility, and tonicity. The use of such terms as the digestive power by Grimaud and the *force de situation fixe* by Barthez were erroneous and confusing. The persons who used them were merely confusing the properties of life with its results, Bichat contended.[40]

Having considered the subject of gradual death due to ageing, Bichat turned in the second part of *La vie et la mort* to the study of violent death from injury. He examined, as we saw in the preceding chapter, the consequences of injury to the heart, the lungs, and the brain, claiming that every sudden death begins by interruption of the circulation, the respiration, or brain activity. While the organic life may exist both before and after the animal life, the latter does not persist even for a moment once the former is extinguished. He chose to study those three specific organs because he believed that the two lives interact by way of the relationship among them. Each of them is essential to the other two, "and since they constitute the three centres in which all the secondary phenomena of the two lives are terminated, whenever they cease to act, the phenomena which depend upon them must cease also and general death ensue."

By describing the three organs as centres, Bichat was reviving a theme with a long history. He admitted as much, saying that physiologists have long been acquainted with the importance of a triple focus of life.[41] To realize just how long, we need only remind ourselves of the tenacious Greek idea of three souls, or pneumata, in which life

[39] Bichat, *Anatomie générale*, pp. cv–cvi.
[40] Bichat, *La vie et la mort*, pp. 121–129.
[41] Ibid., pp. 191–196.

forces were presumed to reside. More recently, Sauvages stated that one of life's essential features is the interaction of the heart and lungs with the soul, which of course, resides in the brain. Bordeu located various centres of sensibility and claimed that the "brain, the heart, and the stomach are . . . the triumvirate, the triple support of life . . . they are the three principal centres from which consciousness and motion flow and to which they return after having circulated."[42] In all these cases, it was taken as axiomatic that forces are diffused from a central source. The tenacity of that belief in a trinity of sources or foci is possibly the most interesting puzzle of all.

The observations concerning violent death were based, as we have observed, upon a multitude of careful and often intricate experiments. The work demonstrates very well how fundamental the animal-organic division was to Bichat's system. Countless observations and experiments were interpreted in such a way as to make them correspond to the demands that that division imposed.[43] Bichat's physiological system, which integrated an antimechanist viewpoint, the animal-organic division, and the sensibility and contractility of bodily parts, produced in total a rather satisfying system. Much about life and the body seemed to be explained in *La vie et la mort*. The book appeared to take account of a very great many themes that ran through the preceding century of work and speculation. In it, Bichat combined the views of Haller, Whytt, Bordeu, Barthez, Grimaud, Buffon, and others. It was compelling material and remarkably persuasive.

Indeed, Bichat's successors found that to take issue with his notions and to set off in a different conceptual and theoretical direction was not always easy. Magendie was one of the first to study and then to dismiss the animal-organic division in print. He pointed out, sensibly enough when we look back on it, that it tended to separate organs and phenomena that are, in fact, intimately connected. According to Bichat, for example, the muscular apparatus that carries food from the mouth to the oesophagus belongs to the animal life, whereas the rest of the gastrointestinal tract belongs to the organic life. Yet all work to a common end. The division, therefore, must be arbitrary and deceptive. We can only concur with Magendie. Bichat's arguments on behalf of the absolute separation of the laws and principles governing life from those governing non-life, on the other hand, were somewhat more difficult to counter. Living phenomena are, after all, almost infinitely variable, so that the predictability that is so satisfying to physicists, astronomers, and chemists constantly eludes physicians. It remained for the students of the life sciences to show that the variability was not merely arbitrary, as Bichat believed, and above all, that it did not break physical and chemical laws, as he maintained. Magendie's student Bernard accomplished the task decisively, but not until the middle of the nineteenth century.

While *La vie et la mort* was being composed, Bichat was already involved with an important new work in anatomy. The vital laws were of considerable importance to the tissue theory of bodily structure. Those persons who approved of the tissue work but considered vital theory to be backward were to be disturbed by the union of those two ideas. For Bichat, however, the tissues were of special significance precisely because he saw them as the structures in which vital forces reside and on which they act.

[42] Bordeu, 'Maladies chroniques', op. cit., note 10 above, pp. 829–831.

[43] A good analysis of these experiments is Sutton, op. cit., note 31 above. See also Chapter V.

VII

THE LIVING ELEMENTS

In his *Anatomie générale*, Bichat compared his twenty-one organic elements to the thirty-three elements of chemistry Lavoisier had described in his *Traité élémentaire de chimie* of 1789. The simple bodies of chemistry, he wrote, are caloric, light, hydrogen, oxygen, carbon, nitrogen, phosphorus, and so on. The simple bodies or tissues of anatomy are (1) the cellular membrane, (2) the nerves of the animal life, (3) the nerves of the organic life, (4) arteries, (5) veins, (6) exhalants, (7) absorbents with their glands, (8) bones, (9) medullary tissue, (10) cartilage, (11) fibrous tissue, (12) fibrocartilaginous tissue, (13) muscles of the animal life, (14) muscles of the organic life, (15) mucous membrane, (16) serous membrane, (17) synovial membrane, (18) glands, (19) dermis, (20) epidermis, and (21) hair (pilous tissue).

Since the nineteenth century, physicians and other commentators have tried to distinguish between Bichat's anatomical and his physiological theories. They did so because they were often troubled by what they saw as the incorrect assumptions that underlay the vitalist physiology. In spite of the limitations of tissue anatomy, which was superseded by cellular theory, these same commentators interpreted it as being basically sound and hence progressive and deserving of their attention. In effect, they would have effectively stripped the anatomical theory of its vitalist substructure, which they felt marred it. Even historians of science have not been exempt from that same tendency. Lain Entralgo, for example, was disturbed by the links of the tissue theory with a vitalism that he would have ignored completely had it been possible to do so. As it was, he merely dismissed it as a "stupidly conservative doctrine which seriously detracted from the beauty of the picture."[1] Such a distinction between allegedly good and bad theories, however, prevented Lain Entralgo from putting Bichat's work into its eighteenth-century context, which is clearer to us because of recent important secondary literature on the subject of monist philosophy.[2] Bichat saw himself as heir to the organicist and monist ideas of physiological function which had been vastly developed in the preceding century. He believed his new anatomical theory to be important, above all, because he considered that he had found the site of the vital forces in the tissues themselves. Vitalism and tissue anatomy were inextricably bound together for Bichat, no more to be separated than the sides of a coin. One cannot truly understand the historical importance of the tissue theory unless one is first aware of its vitalist foundation.

As with many philosophical and scientific ideas, the notion that there exist smallest units of living matter, a kind of biological version of the corpuscular theory, can be

[1] Pedro Lain Entralgo, 'Sensualism and vitalism in Bichat's "Anatomie générale" ', *J. Hist. Med.*, 1948, **3**: 47–64.

[2] In connexion with the monist philosophy of the life sciences, for example, it is worth consulting Walter Pagel, 'The religious and philosophical aspects of Van Helmont's science and medicine', *Bull. Hist. Med.*, 1944, Supplement no. 2. Jacques Roger also deals with aspects of it in his *Les sciences de la vie dans la pensée française du XVIIIe siècle*, Paris, Colin, 1963, pp. 98–103, 585–682.

traced back at least to Aristotle, who made a number of references to simple and to composite parts of animals. In *De partibus animalium*, he went so far as to distinguish "three degrees of composition" in nature. The first was based upon the four elementary material particles of the Greeks. "The second degree of composition is that by which the homogeneous parts of animals, such as bone, flesh and the like, are constituted out of these primary substances. The third and last stage is the composition which forms the heterogeneous parts such as the face, the hand and the rest."[3] Galen wrote a short treatise on the subject of the similar or simple parts in which he commented that primary elements which compose the organic parts are such things as skin, cartilage, bone, various fibres, fat, and so on. Such structures as muscles, arteries, veins, and nerves, because they incorporate various membranes into their structures, are no longer simple.[4] This notion of similar parts continued to have an important place in the scholastic and neoscholastic views of the body, serving as a basis for teaching and exposition in anatomy.

The resemblance between those classical concepts of the simple or homogeneous bodily parts and Bichat's notion of the tissues is merely superficial, of course. It fell to Haller to frame the idea in a form that undoubtedly influenced Bichat, who paid him a considerable and unusual compliment when he advised his readers to pay "homage to his memory by following the route which he traced for us".[5] We have already seen that Bichat imitated certain of Haller's experimental techniques. In his investigation of sensibility and irritability, Haller had examined not only organs but also parts of organs. He differentiated between "simple" parts and "composite" ones. In the former category he included the nerves, arteries, veins, smaller vessels, membranes, muscular fibres, tendinous fibres, ligaments, bone, cartilage, and cellular tissue; the latter are the muscles, tendons, ligaments, viscera, glands, great reservoirs, excretory ducts, and larger blood vessels. Haller did not achieve anything approaching the tissue notion, for he did not conceive of the parts as being specific, distinct physiological and anatomical components. Nevertheless, he distinguished between the sensibility of bone marrow and that of bone; between that of dura mater and that of pia mater; and so on. He also assumed that structures whose composition is basically similar must possess similar qualities and properties. For example, because such membranes as the stomach, womb, intestines, bladder, ureters, and vagina are of the same basic nature as skin, they must be sensible like it.[6]

As we have seen, when Bichat assumed that the forces he attributed to the organic life must exist even though one cannot observe them directly, he was essentially imitating Glisson and the Montpellier vitalists. When he considered the animal life, he tested for its forces just as Haller had looked for sensibility and irritability, albeit keeping in mind the objections of Whytt, who showed that even apparently insensible

[3] Aristotle, *De partibus animalium*, bk. II, ch. 647^{b}, 10–29. Quoted from *The works of Aristotle*, 12 vols., ed. by Sir David Ross, New York, Oxford University Press, 1967.

[4] Owsei Temkin, *Galenism*, Ithaca, N.Y., Cornell University Press, 1973, pp. 12–13; G. Strohmaier, *Galen über die Verschiedenheit der homoiomeren Körperteile*, Berlin DDR, Akademie-Verlag, 1970.

[5] Xavier Bichat, *Traité d'anatomie descriptive*, 5 vols., Paris, Brosson Gabon, 1801–03, vol. 3, p. vii.

[6] Albrecht von Haller, 'A dissertation on the sensible and irritable parts of animals', (London, J. Nourse, 1755), a contemporary translation with an introduction by Owsei Temkin, *Bull. Hist. Med.*, 1936, **4**: 652–699.

bodily parts may become painful. Many of Bichat's conclusions coincided with those of Haller. He reported, for example, that nerves are the organs of animal sensibility, so that all those organs possessing that vital property do so because they have a nerve supply. Animal contractility belongs primarily to the muscles of the animal life. Nerves and muscles lack animal contractility and animal sensibility, respectively. In all these cases, he was merely echoing Haller's notions. In very many respects then, although it was conceptually more complex, the *Anatomie générale* was a kind of extrapolation of Haller's work on the subjects of sensibility and irritability. While the original idea for the *Traité des membranes* came from the work of Pinel, it is still probably true that the mature tissue theory would have been impossible had he not had Haller's example.

Methodologically, of course, the tissue theory was indebted also to sensationalist notions of analysis, which held that the proper way to learn something about an object is to study its component parts.[7] According to Bichat, it is thus that one arrives at a Newtonian "simplicity of causes allied to a multiplicity of effects". Indeed, the *Anatomie générale* was a triumph for ideology. Accordingly, Bichat treated the tissues as living elements, the smallest units into which one can subdivide the organism. Like Aristotle, he would have allowed that these elements can be broken down in turn into those of chemistry. But that is the process of putrefaction, which commences only when life has left the body and the living structures are breaking down. "As in chemistry," he wrote, "the simple [living] substances do not vary though the compounds they unite to form may do so." Nervous tissue, for example, is a membrane in the retina but arranged as cords in the nerves; fibrous tissue is arranged as fasciculi in ligaments, but it is a membrane in the fasciae. Whatever a tissue's form, however, its response to the action of chemical and physical agents is constant. On that assumption, Bichat subjected the bodily parts to various procedures and reagents as well as to dissection. "I have examined every tissue under the influence of caloric, air, water, acids, alkalis, neutral salts, etc. Desiccation, putrefaction, maceration, concretion, etc." Any two parts, wherever they occur in the body, that appeared to behave the same way in response to these various treatments were classified together in one category.

Bichat observed thereby that the muscles of the organic life are more resistant than those of the animal life to maceration, boiling, and putrefaction. Veins are observed to putrefy more readily than arteries. The nervous system of the animal life exposed to acid first undergoes a hardening of its coat, and then a softening. The nervous system

[7] The derivation of the tissue theory from "undeniably sensualist stock" is the thesis of Lain Entralgo, op. cit., note 1 above. The debt of Bichat to Pinel for the tissue theory has been forcefully disputed by Othmar Keel in a number of articles including: 'La pathologie tissulaire de John Hunter', *Gesnerus*, 1980, **37**: 47–61; 'John Hunter et Xavier Bichat: les rapports de leurs travaux en pathologie tissulaire', *Congresso Internacional de Historia de la Medicina. 31 agosto – 6 septembre 1980. Actas*. Barcelona, Acadèmia de Ciéncies Mèdiques de Catalunya i de Balears, 1981, pp. 535–549; 'Les conditions de la décomposition "Analytique" de l'organisme: Haller, Hunter, Bichat', *Études Philosophiques*, 1982, no. 1, 37–62. Keel argues in various contexts that the idea of tissues is implicit in the work of John Hunter and in the work of other persons with whose ideas Bichat must have been familiar even if he did not acknowledge them. He names other persons whose works, predating Pinel's, implied the concept of fundamental tissues. It was apparently the proverbial idea whose time had come. Bichat remains, however, the man who explicitly stated the concept.

of the organic life responds similarly but more slowly. And so on. These are examples *par excellence* of Cabanis' analysis by decomposition.

This alleged unity of physical and physiological properties had important implications for the study of disease. Mindful of Pinel's work, Bichat wrote that if each tissue is unique in health, it must be so in disease also. Diseases normally affect only tissues, spreading to entire organs only if unchecked. Nothing is more rare, he wrote, than affections of the mass of the brain, though it is common to find an affection of its arachnoid tunic; often one eye membrane is affected while others remain normal; in convulsions or paralysis of larynx muscles, mucous surfaces are not affected; in catarrhs, the mucous surface is specifically involved. These are but a few of the examples he offered. It is necessary for the physician, therefore, to study alterations of cellular, arterial, venous, nervous, and other systems rather than diseases of organs and regions. Anatomical observation becomes more important than ever. Twenty years at the sickbed observing diseases of the heart, lungs, and gastric viscera will produce only a confusion of symptoms, which will quickly be dispelled if but a few bodies are opened.[8]

Bichat died shortly after he finished teaching a course in pathological anatomy. We know something of its' content, because the lecture notes of one of his students have been preserved and published as *Anatomie pathologique: dernier cours de Xavier Bichat*. The course was divided into sections treating, consecutively, the afflictions of the serous system, mucous system, cellular system, lungs, glands, skin, muscles of the organic life, arteries, veins, nerves, absorbent system, fibrous system, synovial system, cartilage, medullary system, bones, hair, and epidermis. Certain tissues, including those of the muscles of the animal life, appear not to have been considered in a separate category, but it may simply be that the student's notes are incomplete. According to the notes, for example, Bichat taught that serous membranes are subject to acute and chronic inflammations, rashes, whitish spots, occasional membrane ossification, and various sympathetic effects. In subsections, he treated the affections of the pleura, the pericardium, the peritoneum, the vaginal tunic, and the arachnoid membrane.[9] In *La vie et la mort,* Bichat defined disease as an alteration of certain vital properties, not so much a state "contrary to nature" as a modification of a normal condition.[10] In the *Anatomie pathologique*, we read that "the most active organs are most subject to illness."[11] Consistent with that principle, Bichat found that whereas the mucous membrane is frequently painful or inflamed, the hair is subject only to one rare hereditary malady, the "plique polonaise", which presumably dissolves the largely passive pilous tissue. The epidermis, which enjoys the least

[8] Xavier Bichat, *Anatomie générale appliquée à la physiologie et à la médecine*, 4 vols., Paris, Brosson, Gabon, 1801, vol. 1, pp. lxxxv–xcix. Hereinafter cited as *Anatomie générale*. Michel Foucault, *The birth of the clinic*, trans. by A. M. Sheridan, London, Tavistock, 1973, has much to say on the contribution of Bichat's theories to medical practice and teaching in post-revolutionary France, including Bichat's perception of disease and its relationship to the bodily parts.

[9] Xavier Bichat, *Anatomie pathologique: dernier cours de Xavier Bichat*, based on notes taken by P. A. Béclard, Paris, Baillière, 1852, pp. 38–74. Hereinafter cited as *Anatomie pathologique*.

[10] P. Huard observed that, for Bichat, there were no limits between normal and pathological. The notion of the tissue appeared to him to be probably more important in the ill person than in the healthy one. P. Huard, 'Bichat anatomiste', *Hist. Sci. méd.*, 1972, **6**: 98–106.

[11] Bichat, *Anatomie pathologique*, p. 105.

vitality of all, is subject only to corns, which, in themselves, are entirely painless.

In the *Anatomie pathologique*, Bichat's ambiguity about the role of the blood and other fluids showed up again. He told his students that his words about the nature of illness pertain only to solids, even though fluids are unquestionably altered in certain circumstances. Indeed, fluids such as those of cysts and hydatids are produced only in illness. Insofar as he considered them at all, it seems that Bichat interpreted such effects as merely secondary in nature, perhaps having to do with the fluids' partial animalization.[12]

It follows from his views on illness that Bichat believed that remedies act to restore vital forces from an altered state to their proper form and level. Each of the five vital properties ought to have its own class of appropriate remedies.[13] This was not further developed, however. Buisson informs us that at the time of his death, Bichat had begun work on a *materia medica*. What we know of it is contained in a notebook belonging to L. N. Jusserandot and preserved in the Zurich Medical History Institute. It suggests that Bichat was strongly urging a new classification of medicines based upon their activity. Accordingly, he is said to have distinguished those drugs which act on fluids from those which act on solids, all the while making a multitude of observations on the wards of the Hôtel-Dieu.[14]

There are no fluids, only solids, among Bichat's twenty-one tissues. Some vitalists, among them Barthez, believed that the vital principle exists in both the fluid and solid parts of the body. It is that force, he claimed, which causes the blood to circulate.[15] Bichat did not agree with Barthez in that instance. It is possible that his solidist viewpoint was a kind of byproduct of his basic medical education. A surgeon, after all, necessarily treats only the solid parts of the body, and those only locally. Never having considered fluids to be a locus of treatable lesions, Bichat was not inclined to allow them the same status as the solid parts when it came to anatomical classification. Whether or not this accounts for the special status of fluids, he believed that blood is a kind of reservoir whose composition changes constantly. In general, fluids act as excitants in the parts through which they pass. The blood, for example, stimulates the contraction of the heart. But the same fluids lack the capacity to move on their own or to experience sensation. That is to say, they lack vital properties.

Although he contended that fluids are not the carriers of living forces, Bichat was not prepared to go so far as to claim that they are entirely inert. They must possess, he thought, some intermediate status or quantity of life, if only because they contain within them particles that have been expelled from, or are about to become incorporated into, the animated solids. This problem for animalization or vitalization, however, was largely a mystery for Bichat, as he admitted:

> The alimentary mass is less animalized than the chyle, the chyle less than the blood, etc. It would undoubtedly be a very interesting subject of inquiry to determine how particles hitherto devoid of animal

[12] Ibid., pp. 16–22.

[13] Bichat, *Anatomie générale*, vol. 1, pp. xl–lii.

[14] Mathieu-François Buisson, 'Précis historique sur Marie-François Bichat', in Bichat, op. cit., note 5 above, vol. 3, p. xxv. The existence of the notebook is related in Erwin H. Ackerknecht, *Medicine at the Paris Hospital 1794–1848*, Baltimore, Md., Johns Hopkins University Press, 1967, p. 131.

[15] Paul-Joseph Barthez, *Nouveaux éléments de la science de l'homme*, 2 vols., Montpellier. J. Martel ainé, 1778, vol. 1, pp. 101–117.

properties and enjoying only physical ones should impregnate themselves by degrees with the rudiments of the former .... To say what that vitality of fluids is, is evidently impossible; but its existence is nevertheless real .... Let us observe, in effect, that from the moment the principle of life forsakes the fluids they verge on putrefaction and are decomposed like the solids when deprived of their vital powers.[16]

La Mettrie and especially Diderot had solved that problem, to their satisfaction at least, when they postulated that sensibility is a universal property of matter released under particular forms of organization. But Bichat's assumptions would not permit that solution. His own notions concerning natural forces in general and vital forces in particular had been shaped very strongly by what Barthez had written on the same subject. Vital forces, they both said, are somehow superimposed on inert matter and supersede its own inherent physical forces. The big question, never tackled in an even remotely satisfactory way, was just how such an imposition could take place. The question of vitality came up, therefore, in connexion with the status of the fluids among the tissues. As the quotation above shows, Bichat's speculation was entirely vague, raising more questions that he was inclined to try to answer.

In his *Anatomie générale*, Bichat divided the twenty-one tissues into two major groups according to the way in which they are distributed in the body. The cellular, nervous, vascular, exhalant, and absorbent tissues are dispersed throughout every bodily structure in such a way that if all other material were to be dissolved, they would form an outline of every organ. Their function is necessary for the continued existence of every part of the body. The second group of tissues includes bone, cartilage, tendons, muscles, and the mucous and serous systems. These exist only in certain limited parts of the body, where they perform a more restricted or specialized function than the tissues of the first group.[17]

The *Anatomie générale* is divided into sections labelled "cellular systems", "vascular system of red blood", and so on, and not, as we might perhaps expect, "cellular tissue", and "arterial tissue". Bichat treated systems rather than tissues, because he was interested not only in the form and properties of a particular tissue, but also in its distribution, organization, development, and interrelationships with other tissues and systems. Each of the systems associated with the two muscular systems, for example, is made up not only of muscle tissue but also of the blood vessels, absorbents, secretory vessels, nerves, and other tissues that support, nourish, and generally integrate it into the body as a whole. This implies a basic reluctance on Bichat's part to create a rigid separation between the study of anatomy and that of physiology. An anatomical unit makes sense only in the larger context of the complete organism. It is a logical approach, especially for a vitalist, but it also created certain problems, as I shall try to show.

We have already had occasion to observe how Bichat dealt with the cellular tissue, which is the most abundant of all the parts of the body. This tissue had already undergone considerable discussion before he turned his attention to it. Mention of a "cellular membrane" had been made by both Ruysch and Boerhaave. Haller devoted an entire chapter to it in his *First lines of physiology*, describing it as a supportive structure containing fatty deposits, forming membranes, and acting as a base for

[16] Bichat, *Anatomie générale*, vol. 1, pp. lxi–lxxii.
[17] Ibid., pp. 1–10.

many structures. Perhaps the most extensive work before Bichat's own was the treatment accorded the tissue by Bordeu in his *Recherches sur le tissu muqueux ou l'organe cellulaire.* Bordeu described mucous or cellular tissue as the most extensive part of the body. It nourishes all the organs, forms their base, and connects them one to another. It is the seat of several illnesses and of many phenomena of the animal economy. Under the microscope, Bordeu observed many small translucent bodies or "cells" linked together. Leeuwenhoek had remarked earlier that these cells are the same size in a flea as in a cow. Consequently, Bordeu considered it possible that these cells are somehow the basic material of the animal body.[18] The mucous tissue forms a great sac underneath the epidermis, which is subdivided into one part in the head and neck, one part in the thoracic region, and one in the pelvic area. Throughout the body, it is constricted along the median line of the body to form the *raphé générale.* It is clear enough, therefore, if we compare Bichat's basic treatment of the tissue to that of Bordeu, that Bichat owed something to his predecessor at least with respect to the form, distribution, and function of the tissue. Displaying his usual lack of grace when he referred to Bordeu's work, however, Bichat merely dismissed his *Recherches sur le tissu muqueux* as "some very vague ideas on the subject of the tissue peculiar to the organization of the cellular system which were not even supported by experiment."[19]

When he undertook to examine nervous tissue, Bichat found that it seems to subdivide naturally into two types in accordance with the animal-organic division. The nervous tissue of the animal life has its centre in the brain. It receives external sensations and activates voluntary muscles. That is to say, it controls the muscles of the animal life. The nervous system of the organic life has many centres in the ganglia of the great sympathetic nerve that travels alongside the spinal cord. Its nerves are distributed to the organs of the viscera. Neither system, however, is strictly confined to the organs of its own life, for cerebral nerves send branches to glands and to involuntary muscles, while the ganglia send branches to some voluntary muscles.[20] This admission is interesting, for conceivably it might have suggested to Bichat that the animal-organic division is a less rigid and integral a part of nature than he had once supposed. His addiction to it, however, would not allow him to be deflected.

The most abundant – in fact, virtually the only – property of the nerves of the animal life is animal sensibility, which they transmit from the bodily parts to the brain. Laid bare and excited, they cause the animal much pain. These same nerves, as Haller had long since declared, possess absolutely no animal contractility. Taking no active part in secretion, exhalation, and other internal functions, the cerebral nerves possess few organic properties. Nor are they well endowed with tissue properties, as nerves are rarely stretched.[21]

Bichat described the ganglia or anastomoses, which belong to the nervous system of the organic life, as insulated nerve centres, each one of which functions as a little brain. He saw the great sympathetic nerve as merely a cord which provides a series of

[18] Théophile de Bordeu, 'Recherches sur le tissu muqueux ou l'organe cellulaire', in *Oeuvres complètes*, 2 vols., Paris, Caille et Ravier, 1818, vol. 2, pp. 735–740.

[19] Bichat, *Anatomie générale*, vol. 1, p. 64.

[20] Ibid., pp. 115–118.

[21] Ibid., pp. 125–212.

communications between the numerous nervous systems located one above the other. Any communication between these miniature brains or ganglia is merely accessory. Unlike the regular and symmetrical nerves of the animal life, those of the abdomen, the heart, and other organs of the organic life are irregular as are their ganglia.[22]

The properties and even the functions of organic nerves gave Bichat more difficulty than had those of the animal nerves. He found no animal contractility and only a little animal sensibility in them. Though they affect the sensible organic contractility of the heart and intestines, they do not control it, for cutting them does not annihilate that vital property. In effect, Bichat could find no particular use for these nerves in spite of the abundance in the body. Bordeu had linked the activity of glands and of other viscera to their respective nerve supplies. Had Bichat been prepared to do so, his problem of nervous function would surely have been simpler. He maintained, however, that by their very nature, the organic properties are confined to their respective organs or tissues. Hence, they must exist apart from nerves. This conviction was at the root of his dismissal of Bordeu's evidence concerning the very critical nervous role in glandular activity.[23] The result, however, was confusion for Bichat. Indeed, his treatment of this part of the nervous system was undoubtedly the least satisfactory of all the sections of the *Anatomie générale*.

Bichat's treatment of the parts of the circulatory system again reveals the strength of certain prejudices and presumptions in his work. He distinguished between arterial and venous tissue as thousands had done before him. Nevertheless, he divided his discussion in the *Anatomie générale* into that having to do with the "vascular tissue of the red blood" and the "vascular tissue of the black blood". It was a curious pair of categories, especially for an anatomical work, for it appeared to be a remarkably awkward way of disregarding an obvious anatomical distinction. In its place, Bichat took account primarily of the quality of the fluid that circulates in the various parts of the system. That approach seems to be inconsistent not only with his views about the nature of tissues but also with his belief that the blood is basically an inert mixture of elements. The red blood system, Bichat wrote, originates in the lungs, where blood acquires a colouring principle from the air. It includes the pulmonary veins, the left heart, and the arterial system of the trunk. In the trunk, the blood loses its colouring principle and the resultant vermilion hue. The black blood system includes the venous system of the trunk, the right heart, and the pulmonary arteries. While he admitted the confusion of two types of vessels, he pointed out that his division is logical if one considers the function of the blood in the body. The red blood circulation furnishes the body with the material it needs for secretion, exhalation, and nutrition. The black blood system, on the other hand, is a kind of general reservoir receiving discarded lymph, serous exhalants, and various nutritive wastes. Here above all in this work, Bichat's instincts as an anatomist were deflected by a preoccupation with physiology, which caused him to lose sight of his own precise definition of the word "tissue".[24]

Denying the status of a separate tissue to capillaries, Bichat treated them as the focus of yet two other tissues, the exhalants and the absorbents. He nevertheless

[22] Ibid., pp. 213–218.
[23] Ibid., pp. 220–244.
[24] Ibid., vol. 2, pp. 245–468.

devoted many pages to an enlightening discussion of the place of the capillaries in the animal economy. It is precisely in the capillary vessels of the trunk, he wrote, that red blood is transformed into black blood; in the capillaries of the lungs, black blood takes on a vermilion hue. The system is the focus of such important organic functions as secretion, nutrition, absorption, exhalation, and heat production. The minute canals are the seat of various inflammations. Although the lower classes of animals frequently lack a heart, they all possess the basic and fundamental capillary circulation. It is a kind of link, Bichat speculated, between plants and animals.[25]

Organic sensibility and the organic contractility that inevitably accompanies it are the dominant vital properties in all those tissue systems which act in one way or another through the capillary system.[26] Being the instruments of nutrition, absorption, and so on, capillaries actively become part of every organ of the body. While the capillaries of the muscles, spleen, pituitary, and certain parts of the mucous surface contain only blood, those of the tendons, cartilage, hair, and certain ligaments have no blood at all. Bichat ascribed the separation of the various fluids in this vast and interconnected system to the ubiquitous and highly specific organic sensibility of the various bodily parts:

> It depends entirely on the connection between the organic sensibility of each part of the capillary system and the fluid which it contains . . . . Why does the trachea admit air and repulse all other fluid? All this has to do with the fact that each part, each portion of the organ, each organic molecule has its own type of sensibility so to speak which has a rapport with only one substance and repulses all others.

This explanation of organic sensibility is unmistakably linked to Bordeu's explanation of the selective secretion of the particles from the blood by glands owing to their unique and specific sensibilities. Bichat admitted as much, commenting that "however slightly the phenomena of the capillary system are examined, the facts which Bordeu first recognized will be easily observed".[27]

When he located heat production in the capillaries, Bichat was tackling a phenomenon that has baffled philosophers and physicians for centuries. He observed that some parts of the body are warmer than others. While this is more striking in the case of a local inflammation, he went on, it occurs even in complete health, so that the general temperature of the animal body arises from the combined individual temperatures of many parts. Believing, like Lavoisier, that heat or caloric is a material element, Bichat wrote that blood absorbs it from food and from respired material in the capillary system. Each system has a unique level of heat simply because its secretion, like that of other circulating particles, depends upon the system's specific *insensible organic sensibility:* "Each system has its peculiar mode of heat production just as each gland has its peculiar mode of secretion; each exhaling surface its peculiar mode of exhalation; each tissue its peculiar mode of nutrition and all this directly proceeds from the modifications of the vital properties in each part."[28]

In view of their functions, it follows that the tissues of exhalation and of absorption are distributed throughout all parts of the body. Bichat wrote that exhalation, like

[25] Ibid., pp. 469–470.
[26] Ibid., pp. 487–504.
[27] Ibid., pp. 591–602. The capillary system is discussed on pp. 470–548.
[28] Ibid., pp. 520–536.

secretion, is a process whereby liquids are separated from the blood and poured over various surfaces. While secretion occurs in glands, however, only the capillary plexus separates arteries from exhalant vessels. Exhaled materials are synovial fluid, fat, serum, mucous, bone marrow, and all nutritive substances. Apparently devoid of all animal properties, the exhalant system is governed exclusively by the *organic sensibility* and the *insensible organic contractility* specific to each system. This specificity ensures that mucous exhalant, for example, is different from the serous one. An exhaled fluid that has served its purpose passes through the lymph glands into the minute capillary vessels and thus into the black blood. What Bichat named the absorbent system was the combination of these glands and their vessels. All the known absorbents unite into two principal trunks, delicate and transparent, which finally empty into the superior vena cava. Like the exhalants, they function simply because they possess appropriate organic properties.[29]

The careful reader may find himself uneasy with these two tissue categories, for, once again, Bichat seems to have confused anatomical divisions with physiological functions. In the introduction to the *Anatomie génèrale*, he strongly insisted that he intended to treat structures as though they belong to a single tissue system if they responded similarly to chemical and physical manipulation. However, the existence of absorbents and exhalants was inferred rather than demonstrated, for they could not be directed, observed, or experimentally manipulated. Bichat admitted as much. According to his own definition of a tissue, the capillary vessels deserve their own category. They contain certain apparatus which governs absorption, exhalation, and so on. Having lost sight of his original goal when he discussed the parts of the circulatory system, however, Bichat was prevented from according the capillaries their due status in the body. To permit such inconsistencies to remain in his work, Bichat must have been submitted to considerable pressure from an impatient and uncritical publisher.

The tissues discussed so far are the most basic ones, without which an animal organism could not exist. Accordingly, they are distributed throughout every part of a body. The remaining ones perform more specialized functions and are, therefore, more localized. The bones, cartilage, fibrous tissue, and animal muscles are destined for locomotion; the organic muscles and the serous and mucous tissues are incorporated into the digestive apparatus; the respiratory, circulatory, and glandular tissues together are responsible for secretion; the cutaneous system of the dermis, epidermis, and pilous tissues constitutes the external sensitive apparatus.

Bony tissue possesses only organic properties, except when its sensibility is raised to an animal level because of caries or other bone disease. Bones are formed, Bichat observed, as calcium phosphate is deposited into the cartilaginous skeleton of the foetus.[30] Not all cartilage turns into bone, however. It is a white, hard, elastic substance organized into tightly interlaced fibres. In an adult, it is found only at the articular ends of movable bones and on the parieties of certain cavities, such as the cartilage of the nasal partition, the ribs, and the larynx.[31]

Inside the bones, one finds medullary tissue. Bichat described it as a fine vascular

[29] Ibid., pp. 549–636.
[30] Ibid., vol. 3, pp. 5–104.
[31] Ibid., pp. 119–144.

interlacing that adheres to the inside of all the bones and serves as the exhalant organ of the medullary juice. It possesses animal sensibility, especially in the long bones, he found, for amputation or the introduction of an instrument into the bone causes severe pain when they touch medullary tissue.[32]

The mucous, serous, and fibrous membranes had been examined by Bichat as early as 1798 and featured in his *Traité des membranes*. We have had occasion to consider these tissue categories briefly earlier. In the *Anatomie générale*, the fibrous category was extended to include ligaments and tendons. In the interim, Bichat found that he could isolate a combined fibrocartilaginous tissue, which he claimed occurs in ears, nostrils, the trachea, eyelids, and at vertebral articulations. Fibrous tissue is the common base of these structures, but more like cartilage, they develop animal sensibility when they become inflamed.[33]

The muscle system, Bichat found, naturally divides into animal and organic tissues. The two categories correspond closely to what we know today as the voluntary and involuntary muscles. The muscle tissue of the animal life is the more extensive of the two. It fills numerous regions and is spread out under the skin. Its muscles are red and disposed in obvious fibre bundles. Although Leeuwenhoek had done microscopic studies of these muscle fibres, Bichat naturally dismissed them as a futile search for "the intimate structure of organs" and hence for inaccessible first causes.[34]

In general, Bichat found that the muscles of the animal life possess a remarkable degree of vitality and function more rapidly than does any other organ or bodily part. They alone possess at least a measure of all the various vital forces. Amputation of such a muscle is painful only when its nerve filaments and not the muscle fibres themselves are being cut. Recall that Haller had claimed that voluntary muscles do not possess sensibility. Bichat claimed, nevertheless, that this particular tissue is endowed with animal sensibility, simply because it is very susceptible to the sensation of weariness. Of all the tissues, only the muscles of the animal life have animal contractility. In this, he was in agreement with Haller. Indeed, it is this vital force that accounts for the unique and important functions this type of muscle performs in the body. The primary cause of voluntary animal muscle activity is the soul. The signal for it is transmitted by the nerves of the animal life from the brain, which is the intermediary between the soul and the nerves just as the nerves are intermediaries between the brain and the muscles.

Laid bare, the muscles of the animal life are observed to demonstrate sensible organic contractility, for they react involuntarily to the direct application of various stimuli or irritants. In cases of mental alienation, delirium, head wounds, or inflammation, muscular contractions become involuntary, for the soul's direction is overpowered by sympathetic signals from other parts of the body. Strong passions such as anger emanating from the organic life can also occasionally triumph over the will. Muscle power varies greatly depending upon whether a particular activity has come about because of the activity of the soul, the sympathies, or merely mechanical agents.[35]

[32] Ibid., pp. 105–118.
[33] Ibid., pp. 145–224.
[34] Ibid., pp. 315–317.
[35] Ibid., pp. 224–339.

The muscle tissue of the organic life participates in the formation of the heart, gastrointestinal tract, bladder, and womb. Except for the heart, its fibres are flat and membranous and often curved, folded, and formed into bags and cylinders. Because the tissue is not subject to the will, it does not weary and is not painful if cut or irritated directly. Under normal conditions, therefore, it possesses no animal properties at all. The predominant muscular property in the organic life is sensible organic contractility, which is excited by blood in the heart, urine in the bladder, food in the gastric organs, and so on. "Each individual muscle is possessed of a degree of organic contractility peculiar to it and upon which certain fluids only in the animal economy can act with regularity."

As we observed earlier, Bichat was troubled by the fact that these organic muscles receive nerves from both the brain and the ganglia. The bladder and rectum possess certain limited animal properties. But Bichat could not understand why, for example, any nerves of the animal life should travel to the organic muscles in the abdominal region. He came to no conclusion, merely remarking that he was not sufficiently acquainted with the influence of the brain and nerves on the muscles.[36]

When he came to consider the glandular tissue, Bichat naturally had to refer to Bordeu's work. Much as Bordeu had, Bichat defined a gland as an organ that separates a certain fluid from the blood and expels it through one or more ducts. In this category, he included the salivaries, lachrymal glands, mammae, liver, pancreas, kidneys, prostate, testes, and mucous glands. He did not, however, include the lymphatics, pineal, thyroid, thymus, and suprarenal bodies, because, he said, they do not possess the required excretory duct. Once again, it would appear that Bichat lost sight here of his anatomical definition of "tissue". He himself pointed out the obvious fact that the texture of the glands varies a great deal. That of the liver, for example, is entirely different from that of the kidney, and both differ substantially from the salivary glands. Bichat's instincts as a physiologist gained the upper hand over those of the anatomist. He found evidence, however, that there is such an anatomical element as glandular tissue. He relates that when he subjected various glands to the effects of drying in the air, putrefaction, heating, boiling, acids, and so on, they responded similarly. They reacted in various ways to maceration, however. The liver, for example, resisted the action of the water better than the kidneys, while the salivary glands broke up almost immediately. But that, he suggested, is largely the consequence of the fat contained in each gland and not of its texture.

Dismissing the microscopic studies of Malpighi and Ruysch, Bichat commended Bordeu for having clearly demonstrated that vital and not mechanical activity is the cause of glandular activity. He disagreed sharply, however, with Bordeu's contention that nerves govern this vital action. He pointed out that glands secrete even in cases in which an organ in which they are situated is paralysed and presumably its nerves are inactive, as, for example, the mucous glands in an inactive bladder. Being altogether uncertain of the role of the nervous tissue in the organic life, as we observed, Bichat preferred to believe that glands function simply because they possess the two basic properties of the organic life:

[36] Ibid., pp. 224–414.

> It is by means of its organic sensibility that the gland secretes the materials proper to it from the mass of the blood. It is by means of its insensible contractility or by its tonic powers that this organ contracts and rises . . . to expel those matters heterogeneous to this secretion . . . it is by means of its peculiar mode of organic sensibility that each living part in the economy thus distinguishes what its functions require . . . . In the fluids approaching the small vessels of this gland, this sensibility is the sentry that warns and insensible contractility the agent that opens or shuts the doors of the organ according to the principles which must be admitted or rejected.[37]

Those words could just as easily have been written by Bordeu.

When he came to consider the skin or dermoid tissue, Bichat described it as a sensitive boundary that establishes a relationship between the body and the external world. In effect, it is the outer surface of the animal life. Its internal surface lies against cellular tissue, which is adjacent in turn to muscles. Skin is composed of a passive corion, of reticular bodies that ramify as small vessels on the skin surface, and of the small sensitive papillae on the external surface of the corion. The papillae are the receptors for the sense of touch, providing one with sensations of mass, heat and cold, humidity and dryness, hardness and softness. They are, therefore, the primary organs of the sensory life.[38] The outermost layer of the skin or the epidermal tissue, on the other hand, has so little vitality that it is almost inorganic. It serves as a kind of semi-organized body, which Bichat described as intermediate between the physical and organic realms of nature.[39]

Finally, the hair or pilous tissue, which arises from the cellular tissue, has the least vitality of all. Because man has the most active external life of all the animals, he has the least hair to lessen his contact with external bodies. Curiously, of all the tissues, the almost inert epidermis and the hair alone can replace themselves. Bichat remarked upon this fact without attempting to explain it.[40]

Bichat's final published work, the *Anatomie descriptive*, was intended to complete the analytic process begun in the *Anatomie générale*. Having by now succeeded in decomposing the organs and structures, he proceeded to the next stage of the total process by, theoretically at least, recomposing the organs and the systems from their parts. Whereas the twenty-one tissues were the object of the *Anatomie générale*, Bichat wrote that it was their various recombinations that concerned him in the *Anatomie descriptive*. It was divided into sections dealing with the "apparatus of the animal life" and the "apparatus of the organic life". The former includes the bones and muscles of locomotion, voice, external sensation, and internal sensation, and the organs of feeling and motion in general. The latter apparatus included digestion, respiration, circulation, absorption, and secretion.

The tissue theory seems to have made considerable impact upon the medical world almost as soon as it appeared. The many flaws in the work were merely those of detail and in themselves insufficient to detract from the virtues of the basic theory. By the time he was writing the *Anatomie descriptive* and lecturing in pathology, Bichat seems to have discarded such awkward categories as the red and black blood systems, and have returned to the far more anatomically sound divisions of arterial and venous

[37] Ibid., pp. 569–639.
[38] Ibid., pp. 640–756.
[39] Ibid., pp. 757–791.
[40] Ibid., pp. 792–828.

tissues. For all the compliments it received in its own time and since, the tissue work nevertheless was superseded within a few decades by the cellular theory, which incorporated it. It is, in a very real sense, tissue theory's distant relative. The tissue theory remains, nevertheless, an incisive concept of some importance for the subsequent development of the life sciences.

VIII

AFTER BICHAT

Less than thirty-one years old at the time of his death, Bichat had managed to acquire a unique position within the Paris school and in the French medical world in general. He was remembered by many students with great affection and his reputation as an anatomist and physiologist was enormous. His work left a mark on clinical teaching and practice that would persist for decades. The members of the Paris hospitals and clinical school quickly absorbed his ideas on the living body, making the *Anatomie descriptive* a textbook for their students. His influence on colleagues and successors lingered particularly in pathological anatomy, which he had begun to teach only shortly before his death, basing it upon his tissue theory. Taking their direction from Bichat, members of the Paris school would add much to the field for the next forty or so years. Among those who admitted their indebtedness to Bichat were René-Théophile-Hyacinthe Laënnec, an anatomist who had worshipped Bichat as a student; P. J. Roux, the disciple and friend who finished the *Anatomie descriptive* and who reputedly kept Bichat's alcohol-preserved head at his side for forty-three years; François-Joseph-Victor Broussais, the inventor of "physiological medicine" and leader of Paris medicine after 1816; Guillaume Dupuytren, the head of clinical studies at the Paris school and a great surgeon; Gaspard-Laurent Bayle, a member of the "pathological-anatomical" school, as well as many others.[1]

Furthermore, Bichat's arguments on behalf of the separation of physiology from the physico-chemical sciences seems to have entrenched itself. Claude Bernard (1813–78) reported, for example, that the Paris medical school was still imbued with the "doctrinal errors" of vitalism nearly forty years after Bichat's death. He remembered being reprimanded early in his career by a Professor Gerdy at the Société Philomathique for questioning the assumption that living nature is infinitely variable and hence fundamentally different from the world of physics.[2] Bernard eschewed many of the notions of the vitalists and especially that about the capriciousness of the organism. One can safely infer, nevertheless, that Bichat's arguments on that subject worried him for decades. Indeed, his most famous work, the *Introduction to the study of experimental medicine* published in 1865, was the result of many years of wrestling with Bichat's ghost concerning the nature of medical science and the methodology appropriate to it.

The challenge to Bichat's assumptions started before that, however, with Bernard's mentor François Magendie (1783–1855) who, in the 1820s, published annotated editions of the *Traité des membranes* and the *Recherches physiologiques sur la vie et la*

[1] The central role of Bichat's anatomical and physiological teaching in post-revolutionary France is assumed by two authors who deal specifically with medical and clinical teaching in the period. All the persons mentioned here are discussed in the context in Erwin H. Ackerknecht, *Medicine at the Paris Hospital, 1794–1848*, Baltimore, Md., Johns Hopkins University Press, 1967; and Michel Foucault, *The birth of the clinic*, trans. by A. M. Sheridan, London, Tavistock, 1973.

[2] Claude Bernard, *Lectures on the phenomena of life common to animals and plants*, trans. by Hebbel E. Hoff, Roger Guillemin, and Lucienne Guillemin, Springfield, Ill., Charles C Thomas, 1974, p. 41.

mort which, he remarked, were already classics.[3] Magendie praised Bichat's observational spirit, his experimental genius, and his lucid manner of presenting the facts. Meanwhile, however, he regretted the uncritical acceptance of many of Bichat's hypotheses and wanted to warn students against them. In the fourth edition of *La vie et la mort*, for example, he thundered against the notion of the two lives, and objected to the image of the "animal as a plant clothed in external garb of the organs of relation" on the grounds that it tended to isolate parts and functions which work together to achieve particular results. For example, Bichat would have it that the muscular apparatus of the animal life passes a lump of food from the mouth to the oesophagus while that of the organic life moves it through the remainder of the gastrointestinal tract. As early as 1809, Magendie set out his lifelong theoretical position in *Quelques idées générales sur les phénomènes particuliers aux corps vivants*, in which he criticized vital principles, properties, powers, and forces.[4] He particularly objected to Bichat's vital forces, which were unequally distributed and even, in some cases, limited to particular parts of the body. His point was that if it is a vital property, it ought to be general, characterizing life everywhere. He argued that all living phenomena can be explained by two organic characteristics – *nutrition*, which is a process of decomposition and recomposition; and *action*, which is particular to each group of organs such as contraction is to muscles. In effect, Magendie was criticizing the very explanatory framework in which his predecessors, and especially Bichat, had long laboured.

As his objections to the two lives illustrate, Magendie achieved a new focus for physiology, addressing himself not to the properties and functions of organs or of tissues, but to integrated bodily *functions*. For Bichat, the unit of physiology was the anatomical element. But Magendie started not, for example, with the lung and its tissues but with respiration, which was achieved with many organs besides just the lungs. Accordingly, sensibility and contractility ceased to be causal entities as they were for Bichat and other vitalists and were reduced to mere effects.[5]

One need only go to Magendie's *Précis élémentaire de physiologie* of 1825, however, to get the sense that Bichat's ghost still hovered. Having years before dismissed sensible organic contractility and other vital forces as gratuitous, he asserted yet again that they are the "deplorable illusions of modern physiologists" who believe that "in forging a word like *vital principle* or *vital force*, they have done something analogous to discovering universal gravity".[6] He was writing in the present tense of the verbs. Georges Cuvier, the great naturalist, would voice similar objections to the ill-defined and gratuitous vital principles and forces.[7] One cannot help but think

[3] François Magendie, in Xavier Bichat's, *Recherches physiologiques sur la vie et la mort*, 4th ed., Paris, Gabon Libraire, 1822, esp. pp. v–vii, 4–5, 6–7, 15, 19.

[4] François Magendie, 'Some general ideas on the phenomena peculiar to living bodies', in William Randall Albury, 'Experiment and explanation in the physiology of Bichat and Magendie', *Stud. Hist. Biol.*, 1977, **1**: 107–115.

[5] Ibid.

[6] François Magendie, *Précis élémentaire de physiologie*, Paris, Méquignon-Marvis, 1825, 2nd ed., 2 vol., esp. 'Preface', pp. v–xii, and 'Notions préliminaires', pp. 1–[illegible]

[7] See Georges Cuvier, 'Histoire de la classe des sciences mathématiques et physiques', *Mem. Inst. nat. Sci. Arts: Sci. Math. phys.*, 1806, **7**: 1–79. A translation of pp. 76–79 is found in Albury, op. cit., note 4 above, pp. 105–106. See also Georges Cuvier, 'De Barthez, de Médicus, de Desèze, de Cabanis, de Darwin et de leurs ouvrages', *Histoire des sciences naturelles*, 5 vols., Paris, Fortin, 1843, vol. 4, pp. 27–46.

that the words of such men concerning the nature of physiological explanation must have carried considerable authority. In the final analysis, however, this was really a quibble over classifications and over the ontological status of certain natural causes that had been identified to account for phenomena.

Bernard would take on the more fundamental questions having to do with Bichat's notion of two natural sciences. Bernard's recollection, in 1878, of Professor Gerdy's reproof some forty years before, has to do with just that point. The *Introduction to the study of experimental medicine* represented a high point in his career. Recognized as an important document in Bernard's own time, it remains a classic. It contains the arguments that clinch the case on behalf of one set of laws governing all of nature, such that physiology is grounded in physico-chemical principles. He denied absolutely any great metaphysical or epistemological gulf such as that which Bichat alleged separates the science of life from that of non-life.

Conceding that the behaviour of the living organism is, by all appearances, unpredictable, Bernard warned that all is not what it appears to be. The point is not, he argued, that there are two separate categories of sciences but rather two separate environments. Hence Bernard's famous notion of the internal environment (*milieu intérieur*) of the body, which coexists with the external environment outside the body. It is at once a simple and elegant answer to the conundrum with which the vitalists had challenged the mechanists and the reductionists since the seventeenth century. And it is still taught in physiology courses. Bernard wrote about the internal environment as follows:

> If we limit outselves to the survey of the total phenomena visible from without [a living body], we may falsely believe that a force in living bodies violates the physico-chemical laws of the general cosmic environment, just as an untaught man might believe that some special force in a machine, rising in the air or running along the ground, violated the laws of gravitation. Now a living organism is nothing but a wonderful machine endowed with the most marvellous properties and set going by means of the most complex and delicate mechanism .... In experimentation on inorganic bodies, we need to take account of only one environment, the external cosmic environment, while in the higher living animals, at least two environments must be considered, the external or extra-organic environment and the internal or intra-organic environment .... The great difficulties that we must meet in experimentally determining vital phenomena and in applying suitable means to altering them are caused by the complexity involved in the existence of an internal organic environment .... The circulating liquids, the bloodserum and the intra-organic fluids all constitute the internal environment.[8]

Bernard wanted to be rid of vitalism in physiology because he believed it to be a conceptual barrier to experimentation. From the middle of the eighteenth century, the Montpellier school had maintained that observation is the sole reliable source of data about living creatures, a view disseminated by the *Encyclopédie*. As we have seen, however, Bichat performed many experiments. Nevertheless, Bernard insisted that it was Magendie who established experimentation in medicine when he attacked Bichat's vitalism.[9] Now the organism can be subjected to experiment, according to Bernard, because the biological sciences, like the physical ones, are deterministic. This

[8] Claude Bernard, *The introduction to the study of experimental medicine*, trans. by Henry Copley Greene, New York, Schuman, 1949, pp. 63–64.

[9] Discussed by Albury, op. cit., note 4 above. This was also the claim in J. M. D. Olmsted and E. Harris Olmsted, *Claude Bernard and the experimental method in medicine*, Toronto, Abelard-Schuman, 1952, p. 23, where one reads that Magendie was "the pioneer who brought physiology in France back to the experimental method in which it had been established by Harvey in England".

assures that there is a regularity and hence a predictability in nature. Given the same conditions, the same phenomena will always be observed. Living phenomena appear more variable than physical ones only because there are more variables in the internal environment. Without determinism, biology is not a science:

> Absolute determinism exists indeed in every vital phenomenon; hence biological science exists also .... We must therefore have recourse to analytic study of the successive phenomena of life, and make use of the same experimental method which physicists and chemists employ in analysing the phenomena of inorganic bodies. The difficulties which result from the complexity of the phenomena of living bodies arise solely in applying experimentation; for fundamentally the object and principles of the method are always the same.[10]

Finally, the exorcism seems to have been accomplished.

Interestingly enough, neither Magendie nor Bernard nor many others denied the existence of a supra-physical force to account for the apparently purposeful activities of the living body. Two centuries earlier, observation of the development of an embryo had confirmed William Harvey's vitalistic assumptions at the very time when other men were taking his circulatory theory to be supportive of iatromechanism. While protesting that the embryonic changes are in accordance with the "physico-chemical conditions proper to vital phenomena", Bernard permitted himself to speculate about the existence of a "developing organic force", the "guiding idea of the vital evolution", and a "creative vital force".[11] Life, it would seem, is still greater than the sum of its parts, causing the most committed determinist to bow before its complexity and mystery. On the other hand, neither Magendie nor Bernard treated this force or idea as a hypothesis to be tested. Nor did they speculate about its essential nature.

It is perhaps worth remarking that between 1800 and 1865 the amount of light which physics and chemistry could shed on physiological questions had increased enormously, making it easier for Bernard to affirm his arguments. In 1800, Lavoisier's experiments demonstrating the quantitative analogy between respiration and combustion were the single notable application of chemistry and physics to physiology. That demonstration had not been enough, however, to deflect Bichat from his conviction that there is no correspondence between the organic and inorganic realms of nature. Whereas Lavoisier had argued that control of bodily temperature occurs in the lungs, Bichat came nearer the truth by locating it in the capillaries, alleging that the various organs and parts separate caloric from food and air by means of their insensible organic contractility. Bichat took the fact that most animals maintain a temperature different from their surroundings as an affirmation that living and inert nature are separate realms. It was left to Justus von Liebig, a vitalist, to account for animal heat in an entirely satisfactory manner and thus to solve one of the most fundamental and long-standing mysteries surrounding life. By means of painstaking analytical work recorded in his *Animal chemistry* of 1840, he showed that heat is produced by the oxidation or combustion of food.

By the mid-nineteenth century, vital forces were no longer part of the commonplace

[10] For Bernard's views on determinism in physiology, see op. cit., note 2 above, pp. 16–45. The relationship of the notions of Bernard, Bichat, and others to the problems associated with respiration and animal heat is discussed by June Goodfield, *The growth of scientific physiology*, London, Hutchinson, 1960, pp. 135–364. Paul Bert gives Bernard credit for introducing determinism into biology in the 'Introduction' to op. cit., note 8 above, pp. xiii–xix.

[11] Goodfield, op. cit., note 10 above, p. 161.

language of physiology, having given way to physico-chemical images adapted to the organic world. It was not a question of mechanist arguments disproving vitalist ones. It was, rather, that such vitalist convictions as remained had less and less effect upon the experiments or observations of the physicians. In the eighteenth century, the language of animism and especially vitalism addressed the life of the organism itself. Claiming for themselves the mantle of Newtonianism and sound scientific method generally, the vitalists concerned themselves specifically with those features of organisms which distinguish them from non-living matter. They remarked that what had to be addressed was not so much that our glands filter humours like sieves, or that the heart is a pump, or that the vessels are tubes of moving liquid, but that all the parts are acting in response to needs and want, conscious and unconscious, which act to achieve some integrated goal. The living individual is greater than the sum of its parts. Because they addressed the sensation and the motion belonging to the living body, animists and vitalists helped create the science of physiology. In the eighteenth century, its images were necessarily vitalistic because their formulators wanted to have the measure of that which makes life unique. Their iatromechanist predecessors resorted to mechanical images because they yearned for the certitude that they perceived to belong to physics and mechanics.[12]

While it is true, as students and even scientists are apt to affirm, that the nineteenth century reverted to mechanistic explanations of living phenomena, it was not because they finally shook off the last vestiges of a simpler age. Rather, it was because, possessing its own language thanks to the labour of the vitalists, physiology was finally able to adapt a rapidly developing physics and chemistry to its purposes. It was grounded in experiment and in physico-chemical assumptions, as Magendie and Bernard were convinced it must be, but the techniques are special. It was biophysics and biochemistry. The lessons of the Montpellier vitalists, and especially of Bichat, had not been lost after all.

It is occasionally pointed out that the mechanist-vitalist debate surfaces still. While that is true, it has nothing to do with empirical work or with its interpretation. It is rather a transcendental question having to do with whether one believes that an individual, a species, or all of living nature is merely an evolutionary accident, the ultimate product of some chance cosmic collision, or whether some goal or purpose lies behind it all. The scientist in his laboratory does not frame his questions in response to vitalist or mechanist convictions, for they are matters of faith or inclination and not testable hypotheses. Therefore, all scientists have been mechanists in their laboratories for at least the last one and a half centuries.

The tissue theory, as introduced into anatomy in the late eighteenth century, also gave way to advances in physics and chemistry. The product of the method of "analysis" as expounded by Condillac, and flourishing briefly because of a growing interest in pathology, it was superseded by cellular theory. The tissue was displaced in its role as the ultimate unit of life by the cell. Partly, of course, it was because advances in microscopy in the 1830s made identification of cells possible. But it is also true that as sensibility and irritability were demoted from physiological causes into

[12] This is the point basic to François Duchesneau, *La physiologie des lumières: empiricisme, modèles et théories*, The Hague, Nijhoff, 1982.

mere effects, the units which bore them in the former instance were also diminished in status in the body. People still speak of tissues and Bichat's description of their qualities and distribution remains good and thorough anatomy. Nevertheless, tissues are themselves but compounds of living units.

As is usually the case with origins, it is difficult to say just when and where the cellular theory originated. The English botanist Robert Brown observed in 1832 that there are pockets or cells in plants and that every cell contains a nucleus. In 1838, a paper entitled 'Beiträge zur Phytogenesis', appeared in *Müllers Archiv*. It was by Matthias Schleiden, a botanist in Johannes Müller's famous Berlin laboratory, and it described the plant as a community of cells and each individual cell as the "foundation of the vegetable world". Schleiden described Brown's nucleus as a "cytoblast" or kind of regenerative centre of the cell.

In the same laboratory, the zoologist Theodore Schwann undertook to look for cells in all living tissues, thereby expanding Schleiden's work into a general theory about the basis and origin of living phenomena. The credit for clearly defining the cell's character as an independent living unit properly belongs to still another of Müller's students, Rudolf Virchow. His *Cellularpathologie* of 1858 discusses the cell as the basic unit of disease, much as Pinel and Bichat had discussed the membranes and tissues as the seat of illness. Virchow also established the crucial point that all cells spring from pre-existing cells and thus life perpetuates itself. The concept of the cell as a structural unit common to all forms of life was one of those grand unifying notions that was immediately successful because it took account of great complexity with an efficiency of basic explanation.

Until recently, Bichat has tended to be a somewhat confusing and even vague figure in medical history. On the one hand, there has been a conviction that his work is important, as much as anything because his successors, especially in France, had flattered it and responded to it with enthusiasm. In spite of his very short working life, his approach to anatomy and physiology mightily influenced a school of pathology and left its mark on the regenerated post-revolutionary medical institutions of France. His viewpoints and accomplishments, however, had to do with nineteenth-century developments only indirectly. As we have seen, for persons like Magendie and Bernard, his work was a place from which to commence the development of contrary viewpoints. His insights, therefore, enjoyed only a transitory applicability, giving way quickly to new observations and conceptions. What makes his work worth studying for our purposes is that it is so elegant and thorough an integration of the major themes that run through the eighteenth century. Bichat inherited his vitalistic convictions from a long line of investigators who took issue with the simplistic iatromechanism of the early eighteenth century. Flawed though they were shown to be subsequently, his arguments positing variable living phenomena against predictable physico-chemical ones summarized a position to which many assented. Their ontological status aside, the five vital properties very successfully described living behaviour, at least for a time. Armed with his sensationalist methodology, Bichat achieved a remarkably unified and coherent system. Subsequent advances came about partly because Bichat's critics of the nineteenth century were forced to address his persuasive challenges to mechanism and determinism.

INDEX

C

Index

Index

N

O

P